Elementary Introduction to the Lebesgue Integral

Textbooks in Mathematics

Series editors:
Al Boggess and Ken Rosen

A TOUR THROUGH GRAPH THEROY
Karin R. Saoub

TRANSITION TO ANALYSIS WITH PROOF
Steven Krantz

ESSENTIALS OF MATHEMATICAL THINKING
Steven G. Krantz

ELEMENTARY DIFFERENTIAL EQUATIONS
Kenneth Kuttler

A CONCRETE INTRODUCTION TO REAL ANALYSIS, SECOND EDITION
Robert Carlson

MATHEMATICAL MODELING FOR BUSINESS ANALYTICS
William Fox

ELEMENTARY LINEAR ALGEBRA
James R. Kirkwood

APPLIED FUNCTIONAL ANALYSIS, THIRD EDITION
J. Tinsley Oden

AN INTRODUCTION TO NUMBER THEORY WITH CRYPTOGRAPHY, SECOND EDITION
James Kraft

MATHEMATICAL MODELING: BRANCHING BEYOND CALCULUS
Crista Arangala

ELEMENTARY DIFFERENTIAL EQUATIONS, SECOND EDITION
Charles Roberts

ELEMENTARY INTRODUCTION TO THE LEBESGUE INTEGRAL
Steven G. Krantz

Elementary
Introduction to the
Lebesgue Integral

Steven G. Krantz

CRC Press
Taylor & Francis Group
Boca Raton London New York

CRC Press is an imprint of the
Taylor & Francis Group, an **informa** business

CRC Press
Taylor & Francis Group
6000 Broken Sound Parkway NW, Suite 300
Boca Raton, FL 33487-2742

© 2018 by Taylor & Francis Group, LLC
CRC Press is an imprint of Taylor & Francis Group, an Informa business

No claim to original U.S. Government works

Printed on acid-free paper
Version Date: 20180212

International Standard Book Number-13: 978-1-138-48276-0 (Paperback)

Library of Congress Cataloging-in-Publication Data

Names: Krantz, Steven G. (Steven George), 1951- author.
Title: An introduction to the Lebesgue integral / Steven G. Krantz.
Description: Boca Raton : CRC Press, Taylor & Francis Group, 2018. | Includes
bibliographical references and index.
Identifiers: LCCN 2017061661 | ISBN 9781138482760
Subjects: LCSH: Lebesgue integral--Textbooks. | Integrals,
Generalized--Textbooks. | Measure theory--Textbooks.
Classification: LCC QA312 .K7325 2018 | DDC 515/.43--dc23
LC record available at https://lccn.loc.gov/2017061661

Visit the Taylor & Francis Web site at
http://www.taylorandfrancis.com

and the CRC Press Web site at
http://www.crcpress.com

To the memory of Phil Curtis, a great mentor and friend.

Contents

Preface

Going back to the days of Isaac Newton and Gottfried Wilhelm von Leibniz, and even to Newton's teacher, Isaac Barrow, the integral has been a mainstay of mathematical analysis. The derivative is a useful and attractive device, but it is important because of its interaction with the integral. The integral is a device for amalgamating information. As such, it is a powerful and irreplaceable tool.

Without question, the most widely used integral today is the Riemann integral which dates to the mid-nineteenth century. Physicists, engineers, and many mathematical scientists use the Riemann integral comfortably and effectively. It is a very accessible notion of integral, and one with wide applicability. From the point of view of pure mathematics, however, the Riemann integral has definite and specific limitations.

First of all, the collection of functions that are integrable in the Riemannian sense is limited. Even more importantly, the senses in which

$$\lim_{j \to \infty} \int f_j(x)\,dx = \int \lim_{j \to \infty} f_j(x)\,dx \qquad (*)$$

is true in the context of the Riemann integral are limited. It is a fundamental fact that the majority of important theoretical problems in mathematical analysis reduces to an identity of type $(*)$. Questions about convergence of Fourier series, boundedness of integral operators, convergence of solutions of differential equations, the regularity theory for differential equations, and many others all reduce to a consideration of passing the limit under the integral sign. The Lebesgue integral is important primarily because it allows the identity $(*)$ in a rather broad context.

It is important for the budding mathematical scientist to learn about the Lebesgue integral. But the theory is rather subtle. It is quite a bit trickier than the more popular and accessible Riemann integral. Most texts on the Lebesgue theory are pitched to graduate students, and require considerable sophistication of the student. It is important and useful to have a text on the Lebesgue theory that is accessible to bright undergraduates.

This is such a text. Typically a student would take a course from this book *after* having taken undergraduate real analysis. So this would be fodder for the senior year of college. We have endeavored to keep this book brief and pithy. It has plenty of examples, copious exercises, and many figures. The point is to make this rather recondite subject *accessible*.

One thing that we do in this text, to keep the exposition as simple as possible, is to concentrate our focus and attention on the real line. Abstract measure spaces have their place, but for a first go-around the student should concentrate on learning measure theory in the most basic setting. He/she will already be quite familiar with the real numbers, and will therefore be comfortable internalizing the new ideas of measure theory in that context. We do occasionally discuss abstract measure spaces, and we do discuss product measures, but we do so in a very concrete manner.

The book has copious examples and numerous figures and many exercises. We also include a Table of Notation and a Glossary, just to make the book more complete and accessible. We provide solutions to selected exercises.

It is a pleasure to thank my colleagues, Brian Blank, Robert Burckel, Jerry Folland, Richard Rochberg, and Blake Thornton for helpful remarks about this project. The reviewers provided many useful and constructive comments. As always, I thank my editor, Robert Ross, for his enthusiasm and support.

I look forward to feedback from the users of this text.

— Steven G. Krantz
St. Louis, Missouri

1

Introductory Thoughts

1.1   Review of the Riemann Integral

In this book we focus our attention on real-valued functions. We give an occasional nod to complex-valued functions.

Going back to your days of learning calculus, you have been and are now familiar with the Riemann integral. This is the integral that is modeled on *Riemann sums.* Just to review:

Let $[a, b] \subseteq \mathbb{R}$ be a bounded, closed interval and let $f : [a, b] \to \mathbb{R}$ be a continuous function. A *partition* of $[a, b]$ is a sequence of points $\mathcal{P} = \{x_0, x_1, \ldots, x_k\}$ with

$$a = x_0 \leq x_1 \leq x_2 \leq \cdots \leq x_k = b \, .$$

We let $I_j = [x_{j-1}, x_j]$ be the jth interval in the partition, $j = 1, 2, \ldots, k$. Let $\triangle_j = |x_j - x_{j-1}|$ be the length of the jth interval. Define the *mesh* of the partition to be

$$m(\mathcal{P}) = \max_{j=1,\ldots,k} \triangle_j \, .$$

For each j, select a point $\xi_j \in I_j$. Refer to Figure 1.1.

We define the *Riemann sum* of the function f based on the partition $\mathcal{P}$ to be

$$\mathcal{R}_{\mathcal{P}} = \sum_{j=1}^{k} f(\xi_j) \cdot \triangle_j \, .$$

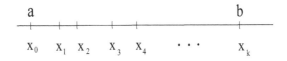

FIGURE 1.1
A partition.

1

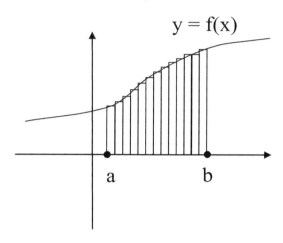

FIGURE 1.2
A Riemann sum.

We think of this sum as giving an approximation to the area under the graph of f and between the left-right limits a and b. See Figure 1.2.

If the limit

$$\lim_{m(\mathcal{P}) \to 0} \mathcal{R}_{\mathcal{P}}$$

exists, then we call the limit the *Riemann integral of the function f on the interval $[a, b]$* and we denote it by

$$\int_a^b f(x)\, dx\,.$$

Now the most fundamental result about the Riemann integral is this:

Theorem 1.1 *Let f be a continuous function on the interval $[a, b]$. Then the Riemann integral*

$$\int_a^b f(x)\, dx$$

exists.

We shall not provide the proof of this theorem here, but instead refer the reader to [4, Ch. 7]

It is not difficult to see that a *piecewise continuous* function will also be integrable. In fact the definitive result about Riemann integrability—see [7]— is the following:

Theorem 1.2 *A function f on a closed interval $[a, b]$ is Riemann integrable if and only if the set of discontinuities of f has measure 0.*

Of course we do not really know yet what "set of measure zero" means; in fact this is one of the big ideas that we are going to learn in this text. But, for the moment, you can take "measure zero" to mean "zero length."

The big feature that differentiates the Lebesgue integral from the Riemann integral is the way that these two theories treat limits. The basic limit theorem for the Riemann integral is

Theorem 1.3 *Let f_j be Riemann integrable functions on a bounded interval $[a, b]$. If $f_j \to f$ uniformly as $j \to \infty$, then*

$$\int_a^b f_j(x)\,dx \to \int_a^b f(x)\,dx$$

as $j \to \infty$.

We now present a couple of examples to suggest the shortcomings of the Riemann integral.

EXAMPLE 1.4 In what follows, if $E \subseteq \mathbb{R}$ is a set, then we let the *characteristic function* χ_E of E be defined by

$$\chi_E(x) = \begin{cases} 1 & \text{if} \quad x \in E, \\ 0 & \text{if} \quad x \notin E. \end{cases}$$

Now define

$$f_j(x) = \chi_{[j,j+1]}.$$

These functions *do not* converge uniformly. So the Riemann theory can say nothing about the sequence of integrals $\int f_j(x)\,dx$. It is nevertheless clear that the indicated integrals converge to 1 as $j \to +\infty$. But, as we shall see below, the Lebesgue theory can. In fact we note here, with reference to Theorem 1.3, that

$$0 = \int \lim_{j \to +\infty} f_j(x)\,dx \neq \lim_{j \to +\infty} \int f_j(x)\,dx = 1.$$

As a second example, let

$$g_j(x) = \chi_{[1,j]}.$$

These functions *do not* converge uniformly. So the Riemann theory can say nothing about the sequence of integrals $\int g_j(x)\,dx$. [It is nevertheless clear that the indicated integrals converge to $+\infty$.] But, as we shall see below, the Lebesgue theory can. Again, with reference to Theorem 1.3, we note that

$$+\infty = \int \chi_{[1,+\infty)}\,dx = \int \lim_{j \to +\infty} f_j(x)\,dx = \lim_{j \to +\infty} \int f_j(x)\,dx$$
$$= \lim_{j \to +\infty} (j - 1) = +\infty.$$

Next, let

$$h_j(x) = \sum_{\ell=1}^{j} \chi_{[0,2^{-\ell}]} \, .$$

Intuitively, the integrals $\int h_j(x)\, dx$ converge to $\sum_{\ell=1}^{\infty} 2^{-\ell} = 1$. But the functions do not converge uniformly so the Riemann integral cannot give this result. We note that, for this example, both sides of the display in Theorem 1.3 equal $\sum_{\ell=1}^{\infty} 2^{-\ell}$.

Now we let

$$m_j(x) = j \cdot \chi_{[j,j+1/j^3]} \, .$$

Although it is intuitively clear that $\int m_j(x)\, dx$ converges as $j \to +\infty$, the Riemann integral will not give this result. With reference to Theorem 1.3, we see that

$$0 = \lim_{j \to +\infty} \int m_j(x)\, dx = \int \lim_{j \to +\infty} m_j(x)\, dx = 0 \, .$$

EXAMPLE 1.5 Refer to the last example for notation.

Let $\{q_j\}$ be an enumeration of the rational numbers. Define

$$f_1(x) = \chi_{\{q_1\}} \, ,$$

$$f_2(x) = \chi_{\{q_1\}} + \chi_{\{q_2\}} \, ,$$

$$f_3(x) = \chi_{\{q_1\}} + \chi_{\{q_2\}} + \chi_{\{q_3\}} \, ,$$

etc. Then the sequence of functions $\{f_j\}$ converges to $f(x) = \chi_{\mathbb{Q}}$. Each of the functions f_j is Riemann integrable by Theorem 1.2. But the function f is *not* Riemann integrable. It will turn out that the function f *is* Lebesgue measurable and Lebesgue integrable.

Let $\widetilde{C}$ be a set constructed like the Cantor ternary set (see the review below and also [4]) but in which intervals of length 5^{-j} are removed at each step. Then $[0,1] \setminus \widetilde{C}$ has length less than 1 and hence $\widetilde{C}$ has positive measure. Consider $f(x) = \chi_{\widetilde{C}}(x)$. Then, by Theorem 1.2, f is *not* Riemann integrable. But it *is* Lebesgue integrable.

1.2   The Idea of the Lebesgue Integral

As you saw in the last section, the Riemann integral is predicated on the idea of breaking up the *domain* of the function f. By contrast, the Lebesgue integral (as we shall see below) is predicated on the idea of breaking up the *range* of the function f.

First we need to discuss the idea of measure. Our dream is to be able to measure the length of any set of reals. Certainly we know how to measure the

FIGURE 1.3
The length of a set.

FIGURE 1.4
First step in the construction of the Cantor set.

length of an interval $[a, b]$. It is of course $b - a$. And we can measure the length of a finite union of intervals

$$S = [a_1, b_1] \cup [a_2, b_2] \cup \cdots \cup [a_k, b_k]$$

with

$$a_1 < b_1 < a_2 < b_2 < \cdots < a_k < b_k.$$

The answer, of course, is

$$\sum_{j=1}^{k} (b_j - a_j).$$

See Figure 1.3.

But what about more complicated sets? For example, what is the length of the Cantor ternary set (see [4])? Refer to Figures 1.4, 1.5. 1.6 to remind yourself how the Cantor ternary set is constructed. Recall that we do so by intersecting

$$
\begin{aligned}
I_0 &= [0, 1] \\
I_1 &= [0, 1/3] \cup [2/3, 1] \\
I_2 &= [0, 1/9] \cup [1/9, 1/3] \cup [2/3, 7/9] \cup [8/9, 1]
\end{aligned}
$$

etc. The Cantor set is uncountable, has length 0, and is perfect.

An idea that goes back to the mid-nineteenth century is this. Suppose that $E \subset \mathbb{R}$ is a closed, bounded set. Then E is contained in a large open interval I. And $I \setminus E$ is an open set, so it is a union of open intervals. We can measure the length of each of those open intervals, and then subtract all those lengths from the length of I. And that will give us the length of E.

0   1

FIGURE 1.5
Second step in the construction of the Cantor set.

FIGURE 1.6
Third step in the construction of the Cantor set.

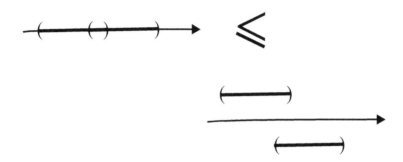

FIGURE 1.7
Subadditivity.

The reasoning in the last paragraph is correct, but it fails to treat a great many sets which we would like to measure. We need a technique for calculating length which treats all sets at once. But it turns out that there is a significant roadblock to this program. In fact *it is impossible to assign a length to every set of real numbers.*

How can this be? In order to answer this question, we should first consider what properties we would want a notion of length to possess. Let $m(S)$ denote the purported length of the set S. We would want m to have these properties:

- $m(S) \geq 0$ for every set S;

- $m(A \cup B) \leq m(A) + m(B)$ (subadditivity—see Figure 1.7);

- If $A_1, A_2, \ldots$ are countably many pairwise disjoint sets then

$$m\left(\bigcup_{j=1}^{\infty} A_j\right) = \sum_{j=1}^{\infty} m(A_j).$$

- If $a \in \mathbb{R}$ then define $\tau_a(S) = \{s + a : s \in S\}$ to be the translation of S by a units. Then we mandate that $m(\tau_a(S)) = m(S)$. In other words, the measure should be *translation invariant*.

We note in passing that the third property here is called *countable additivity*.

EXAMPLE 1.6 We work in the real numbers $\mathbb{R}$. Let $A = [0, 3/4]$ and $B = [1/4, 1]$. Then $A \cup B = [0, 1]$. So

$$
\begin{aligned}
1 &= m(A \cup B) = m([0,1]) \leq m([0, 3/4]) + m([1/4, 1]) = m(A) + m(B) \\
&= \frac{3}{4} + \frac{3}{4} = \frac{3}{2}.
\end{aligned}
$$

We in fact have strict inequality because A and B have nontrivial overlap.

Remark 1.7 It will be common in this book for us to consider a sum of the form

$$
\sum_{j=1}^{\infty} a_j,
$$

where each $a_j = 0$. The value of such an infinite sum, or series, must be 0. For that value is *defined* to be the limit as $N \to +\infty$ of the partial sums

$$
S_N = \sum_{j=1}^{N} a_j,
$$

and each of those partial sums is equal to 0.

It turns out that the following is the case:

Theorem 1.8 (Vitali, 1905) *There is no notion of length m that will satisfy all three of the above bulleted properties for every set of real numbers.*

Proof: We perform this construction on the unit interval $I = [0, 1]$. We do arithmetic on I modulo 1. That is, when we add two numbers together, we subtract off any multiples of 1 to get the answer. Thus

$$
\frac{3}{4} + \frac{3}{4} = \frac{1}{2} \quad , \quad \frac{1}{2} + \frac{5}{8} = \frac{1}{8}
$$

are two examples.

Now define a relation on I by $x \sim y$ if $x - y$ is rational. It is easy to check that this is an equivalence relation. There are uncountably many equivalence classes, and each equivalence class has countably many elements. We use the Axiom of Choice (see [3]) to select one element from each equivalence class. Let S be the set of those selected elements. We claim that S is the nonmeasurable set that we seek.

If q is a nonzero rational number then consider the set $S + q = \{s + q : s \in S\}$. We claim that $S + q$ is disjoint from $S + q'$ for q, q' distinct rational numbers. In fact if $x \in S + q \cap S + q'$, then $x = s + q$ and $x = s' + q'$ for some $s, s' \in S$. But this would mean that $s - s'$ is rational. Hence s and s' are in the same equivalence class. But they are not! The only way that $s - s'$ could be rational is if $s = s'$. But then $q = q'$. Hence our two sets $S + q$ an $S' + q'$ are actually the same.

So $S + q$ and $S + q'$ are disjoint. And these two sets are geometrically identical—one is just a translate of the other.

We know that

$$I = \bigcup_{q \in [0,1] \cap \mathbb{Q}} S + q. \tag{1.8.1}$$

This is just because every element of I has the form $s + q$ for some $s \in S$ and some $q \in \mathbb{Q}$.

Now what is the measure of S? If it is 0, then each $S + q$ also has measure 0 and then it follows from (1.8.1) that I has measure 0. That is of course false. If instead S has positive measure σ then each $S + q$ has positive measure σ. But then it follows from (1.8.1) that I has infinite measure. That, too, is false. So there is no measure that we can logically assign to S.   □

It is interesting that we used the Axiom of Choice to construct the set S above. In fact Solovay [5] has shown that, if you deny the Axiom of Choice, then it is possible to assign a measure to *every* set of reals.

It turns out that we will be able to identify an algebra of sets which we call *measurable*. It is the measurable sets that we can actually measure; other sets we do not attempt to measure. Fortunately, the collection of measurable sets is quite robust and is closed under reasonable mathematical operations. It is a fairly straightforward collection of objects. We shall begin to develop these ideas in the next section.

1.3   Measurable Sets

Throughout this book we use the notation $\equiv$ to mean "is defined to be."

A *σ-algebra* of sets in $\mathbb{R}$ is a collection $\mathcal{X}$ of sets satisfying these axioms:

(a) $\emptyset, \mathbb{R}$ both belong to $\mathcal{X}$;

(b) If $A \in \mathcal{X}$, then $^cA \equiv \mathbb{R} \setminus A$ belongs to $\mathcal{X}$;

(c) If $\{A_j\}$ is a sequence of sets in $\mathcal{X}$, then $\cup_j A_j$ belongs to $\mathcal{X}$.

We call the ordered pair $(\mathbb{R}, \mathcal{X})$ a *measure space* (later on we shall augment this definition). Any set that is an element of $\mathcal{X}$ is called a *measurable set*. We will sometimes consider a measure space with abstract σ-algebra $\mathcal{X}$ on an abstract set X rather than the more specific Borel sets on $\mathbb{R}$ (see below for a discussion of the Borel sets).

We may use de Morgan's laws of logic, together with properties **(b)** and **(c)** of a σ-algebra, to see that the countable intersection of measurable sets is measurable. Namely,

$$\bigcap E_j = {}^c \left(\bigcup {}^c E_j \right).$$

If each E_j is measurable, then the set on the right is measurable because it is formed with complementation and union. Hence the set on the left is measurable.

EXAMPLE 1.9 We now present several examples of σ-algebras.

I. Let $\mathcal{X} = \{$all subsets of $\mathbb{R}\}$. Then it is straightforward to verify properties **(a)**, **(b)**, **(c)** of a σ-algebra.

II. Recall that a set is denumerable if it is either empty or finite or countable. Let $\mathcal{X}$ be those subsets of $\mathbb{R}$ which are either denumerable or have denumerable complement. Then it is easy to check **(a)**, **(b)**, **(c)** of a σ-algebra.

III. For us this will be the most important example of a σ-algebra. Namely, let $\mathcal{B}$ be the σ-algebra generated by the collection of open intervals. That is to say, we are considering all sets that can be obtained by taking **(i)** finite or countable unions of open intervals, **(ii)** finite or countable intersections of open intervals, or **(iii)** finite or countable unions or intersections of sets of types **(i)** or **(ii)**. This σ-algebra is called the *Borel sets*.

IV. This last example is a slight extension of Example **III** that is useful for measure theory. Treat the points $-\infty$ and $+\infty$ as formal objects. If E is a Borel set as in **III**, then set

$$E_1 = E \cup \{-\infty\},$$

$$E_2 = E \cup \{+\infty\},$$

$$E_3 = E \cup \{-\infty, +\infty\}.$$

Now let $\widehat{\mathcal{B}}$ be all Borel sets together with all sets E_1, E_2, E_3 that are obtained from all possible Borel sets E. It is straightforward to check that this new $\widehat{\mathcal{B}}$ is a σ-algebra. We call this σ-algebra the *extended Borel sets*, and we often denote it by $\widehat{\mathcal{B}}$.

It will turn out that the collection of sets to which we can unambiguously assign a length or measure will form a σ-algebra. That σ-algebra will be very closely related to **III** and **IV** in the last example.

Definition 1.10 Let $\mathcal{X}$ be a σ-algebra on $\mathbb{R}$. A function $f : \mathbb{R} \to \mathbb{R}$ is said to be $\mathcal{X}$-*measurable* if, for each real number α, the set

$$\{x \in \mathbb{R} : f(x) > \alpha\} \tag{1.10.1}$$

belongs to $\mathcal{X}$.

Remark 1.11 If S is the nonmeasurable set constructed in the proof of Theorem 1.8, then the function $f(x) = \chi_S$ will *not* be measurable. We will not be able to integrate this function f.

Lemma 1.12 *The following four statements are equivalent for a function* $f : \mathbb{R} \to \mathbb{R}$ *and a σ-algebra $\mathcal{X}$ on a set X.*

(a) *For every $\alpha \in \mathbb{R}$, the set $X_\alpha \equiv \{x \in \mathbb{R} : f(x) > \alpha\}$ belongs to $\mathcal{X}$.*

(b) *For every $\alpha \in \mathbb{R}$, the set $Y_\alpha \equiv \{x \in \mathbb{R} : f(x) \leq \alpha\}$ belongs to $\mathcal{X}$.*

(c) *For every $\alpha \in \mathbb{R}$, the set $Z_\alpha \equiv \{x \in \mathbb{R} : f(x) \geq \alpha\}$ belongs to $\mathcal{X}$.*

(d) *For every $\alpha \in \mathbb{R}$, the set $W_\alpha \equiv \{x \in \mathbb{R} : f(x) < \alpha\}$ belongs to $\mathcal{X}$.*

Proof: Since X_α and Y_α are complementary, statement (a) is equivalent to statement (b). Likewise, statements (c) and (d) are equivalent.

If (a) holds, then $X_{\alpha - 1/j}$ belongs to $\mathcal{X}$ for each positive integer j. Since

$$Z_\alpha = \bigcap_{j=1}^{\infty} X_{\alpha - 1/j}, \qquad (1.12.1)$$

it follows that $Z_\alpha \in \mathcal{X}$. Thus (a) implies (c).

In the same fashion, since

$$X_\alpha = \bigcup_{j=1}^{\infty} Z_{\alpha + 1/j}, \qquad (1.12.2)$$

it follows that (c) implies (a).

A similar argument shows that (b) and (d) are logically equivalent. We conclude that all four statements are logically equivalent. $\square$

EXAMPLE 1.13 We now give several examples of measurable functions.

(1) Let $f(x) \equiv c$ be a constant function. If $\alpha \geq c$, then

$$\{x \in \mathbb{R} : f(x) > \alpha\} = \emptyset.$$

If instead $\alpha < c$, then

$$\{x \in \mathbb{R} : f(x) > \alpha\} = \mathbb{R}.$$

Since both $\emptyset$ and $\mathbb{R}$ are in the σ-algebra $\mathcal{X}$, no matter what $\mathcal{X}$ may be, we see that f is measurable.

(2) Let $\mathcal{B}$ be the Borel sets. Let $f : \mathbb{R} \to \mathbb{R}$ be any continuous function. Then f is measurable because, for any $\alpha \in \mathbb{R}$, the set

$$\{x \in \mathbb{R} : f(x) > \alpha\}$$

is open hence Borel.

(3) Fix a σ-algebra $\mathcal{X}$. Let $E \in \mathcal{X}$. Define the *characteristic function*

$$\chi_E(x) = \begin{cases} 1 & \text{if} & x \in E, \\ 0 & \text{if} & x \notin E. \end{cases} \qquad (1.13.1)$$

Then χ_E is measurable. In fact the set $\{x \in \mathbb{R} : \chi_E(x) > \alpha\}$ is either $\mathbb{R}$, E, or $\emptyset$.

(4) Let S be the nonmeasurable set constructed in the proof of Theorem 1.8. Then the function $f(x) = \chi_S(x)$ is not measurable. This is so because

$$\{x \in \mathbb{R} : f(x) > 1/2\} = S,$$

which is not measurable.

(5) Let $\mathcal{B}$ be the Borel sets. Consider any monotone increasing function $f : \mathbb{R} \to \mathbb{R}$. Let $\alpha \in \mathbb{R}$. Then $\{x \in \mathbb{R} : f(x) > \alpha\}$ is either a halfline of the form $\{x \in \mathbb{R} : x > \gamma\}$ or $\{x \in \mathbb{R} : x \geq \gamma\}$ or the entire line $\mathbb{R}$ or $\emptyset$. Each of these sets is Borel.

It is useful to know that the collection of measurable functions is closed under standard arithmetic operations.

Lemma 1.14 *Fix a σ-algebra $\mathcal{X}$. Let f and g be $\mathcal{X}$-measurable, real-valued functions and let c be a real number. Then each of the functions*

$$cf \ , \ f^2 \ , \ f + g \ , \ f \cdot g \ , \ |f|$$

is measurable.

Proof: For the first result, suppose without loss of generality that $c > 0$. Then, for $\alpha > 0$,

$$\{x \in \mathbb{R} : cf(x) > \alpha\} = \{x \in \mathbb{R} : f(x) > \alpha/c\} \in \mathcal{X}.$$

For the second result, also assume that $\alpha > 0$ (the case $\alpha \leq 0$ is trivial). Then

$$\{x \in \mathbb{R} : f^2(x) > \alpha\}$$
$$= \ \{x \in \mathbb{R} : f(x) > \sqrt{\alpha}\} \cup \{x \in \mathbb{R} : f(x) < -\sqrt{\alpha}\} \in \mathcal{X}.$$

For the third result, and $\alpha > 0$, define the set

$$S_r = \{x \in \mathbb{R} : f(x) > r\} \cap \{x \in \mathbb{R} : g(x) > \alpha - r\}$$

for r a rational number. Obviously $S_r \in \mathcal{X}$ for each r. By considering the cases **(i)** $f(x) > 0$ and $g(x) - \alpha < 0$, **(ii)** $f(x) < 0$ and $g(x) - \alpha > 0$, and **(iii)** $f(x) > 0$ and $g(x) - \alpha > 0$, one may check that

$$\{x \in \mathbb{R} : (f+g)(x) > \alpha\} = \bigcup_{r \text{ rational}} S_r \in \mathcal{X}.$$

Thus $\{x \in \mathbb{R} : (f+g)(x) > \alpha\}$ lies in $\mathcal{X}$, so $f + g$ is measurable.

For the fourth result, observe that

$$f \cdot g = \frac{1}{4}\left[(f+g)^2 - (f-g)^2\right].$$

Thus the measurability of $f \cdot g$ follows from the first three results.

For the fifth assertion, assume as before that $\alpha > 0$. Then

$$\{x \in \mathbb{R} : |f(x)| > \alpha\} = \{x \in \mathbb{R} : f(x) > \alpha\} \cup \{x \in \mathbb{R} : f(x) < -\alpha\} \in \mathcal{X}. \quad \square$$

EXAMPLE 1.15 Let f be a function from $\mathbb{R}$ to $\mathbb{R}$. Define

$$f^+(x) = \max\{f(x), 0\} = \frac{f(x) + |f(x)|}{2}$$

and

$$f^-(x) = \max\{-f(x), 0\} = \frac{|f(x)| - f(x)}{2}.$$

We think of f^+ as the *positive part* of f and f^- as the *negative part* of f. Observe that $f = f^+ - f^-$.

In view of the preceding results, we see immediately that f is measurable if and only if both f^+ and f^- are measurable.

In dealing with sequences of measurable functions, it is often convenient to consider suprema, infima, limsup, liminf, etc. Therefore we want to allow functions to take values in the extended reals (i.e., to take the values $+\infty$ and $-\infty$ as well as the usual real values). We wish to discuss measurability for functions taking values in the extended reals. In what follows, we denote the extended reals by $\widehat{\mathbb{R}}$.

Definition 1.16 Let f be a function from the reals to the extended reals. Let $\widehat{\mathcal{B}}$ be the extended Borel sets. We say that f is $\widehat{\mathcal{B}}$-measurable if, for each real number α, the set

$$\{x \in \mathbb{R} : f(x) > \alpha\}$$

lies in $\widehat{\mathcal{B}}$.

Notice that, if f is a $\widehat{\mathcal{B}}$-measurable function that takes values in the extended reals $\widehat{\mathbb{R}}$, then

$$\{x \in \mathbb{R} : f(x) = +\infty\} = \bigcap_{j=1}^{\infty}\{x \in \mathbb{R} : f(x) > j\}$$

and

$$\{x \in \mathbb{R} : f(x) = -\infty\} = \mathbb{R} \setminus \left[\bigcup_{j=1}^{\infty} \{x \in \mathbb{R} : f(x) > -j\} \right] .$$

Thus both these sets belong to $\widehat{\mathcal{B}}$.

The following somewhat technical lemma is often useful in dealing with extended real-valued functions.

Lemma 1.17 *Let f be an extended real-valued function. Then f is $\widehat{\mathcal{B}}$-measurable if and only if the sets*

$$E = \{x \in \mathbb{R} : f(x) = +\infty\}$$

and

$$F = \{x \in \mathbb{R} : f(x) = -\infty\}$$

belong to $\widehat{\mathcal{B}}$ and the real-valued function f^* defined by

$$f^*(x) = \begin{cases} f(x) & \text{if} \quad x \notin E \cup F, \\ 0 & \text{if} \quad x \in E \cup F. \end{cases}$$

is $\mathcal{B}$-measurable.

Proof: If f is $\widehat{\mathcal{B}}$-measurable, then it has already been shown that E and F belong to $\widehat{\mathcal{B}}$. Let $\alpha \in \mathbb{R}$ and $\alpha \geq 0$. Then

$$\{x \in \mathbb{R} : f^*(x) > \alpha\} = \{x \in \mathbb{R} : f(x) > \alpha\} \setminus E.$$

If instead $\alpha < 0$, then

$$\{x \in \mathbb{R} : f^*(x) > \alpha\} = \{x \in \mathbb{R} : f(x) > \alpha\} \cup F.$$

Thus f^* is $\mathcal{B}$-measurable.

For the converse, assume that $E, F \in \widehat{\mathcal{B}}$ and also that f^* is $\mathcal{B}$-measurable. Let $\alpha \geq 0$. Then

$$\{x \in \mathbb{R} : f(x) > \alpha\} = \{x \in \mathbb{R} : f^*(x) > \alpha\} \cup E.$$

Also, if $\alpha < 0$, then

$$\{x \in \mathbb{R} : f(x) > \alpha\} = \{x \in \mathbb{R} : f^*(x) > \alpha\} \setminus F.$$

Thus f is $\widehat{\mathcal{B}}$-measurable. □

It follows from 1.14 and 1.17 that, if f is extended real-valued and measurable, then the functions

$$cf , \; f^2 , \; |f| , \; f^+ , \; f^-$$

are all measurable.

Remark 1.18 Just a few comments about arithmetic in the extended reals.

By convention, we declare that $0 \cdot (+\infty) = 0 \cdot (-\infty) = 0$. If extended real-valued, measurable functions f and g take the values $+\infty$ and $-\infty$ respectively at a point x or the values $-\infty$ and $+\infty$ respectively at a point x, then the quantity $f(x) + g(x)$ is *not* well defined. But we may declare the value to be zero at such a point, and the resulting function $f + g$ will then be well defined. If f and g *both* take the value $+\infty$ at the point x, then we understand that $(f + g)(x) = +\infty$. Likewise for $-\infty$. Finally, in these last circumstances, $f(x) \cdot g(x) = +\infty$.

We would next like to see that the collection of measurable functions is closed under various limiting operations.

We begin by recalling some definitions.

Definition 1.19 Let $\{x_j\}$ be a sequence of real numbers. For $k = 1, 2, \ldots$, we set

$$y_k = \inf\{x_j : j \in \mathbb{N}, j \geq k\}$$

and

$$z_k = \sup\{x_j : j \in \mathbb{N} : j \geq k\}.$$

Set

$$y = \lim_{k \to \infty} y_k = \sup\{y_k : k \in \mathbb{N}\}$$

and

$$z = \lim_{k \to \infty} z_k = \inf\{z_k : k \in \mathbb{N}\}.$$

Of course y and z are elements of the extended reals. We call y the *limit inferior* or *liminf* of the sequence $\{x_j\}$ and denote it by $\liminf x_j$. We call z the *limit superior* or *limsup* of the sequence $\{x_j\}$ and denote it by $\limsup x_j$.

In fact $\liminf x_j$ is the least of all subsequential limits of $\{x_j\}$ and $\limsup x_j$ is the greatest of all subsequential limits of $\{x_j\}$.

Proposition 1.20 *Let $\{f_j\}$ be a sequence of $\mathcal{X}$-measurable functions. Define*

$$f(x) = \inf_j f_j(x) \quad , \quad g(x) = \sup_j f_j(x) \; ,$$

$$\underline{f}(x) = \liminf_{j \to \infty} f_j(x) \quad , \quad \overline{g}(x) = \limsup_{j \to \infty} f_j(x).$$

Then $f, g, \underline{f}$, and $\overline{g}$ are all $\mathcal{X}$-measurable.

Proof: Let $\alpha \in \mathbb{R}$. Notice that

$$\{x \in \mathbb{R} : f(x) \geq \alpha\} = \bigcap_{j=1}^{\infty} \{x \in \mathbb{R} : f_j(x) \geq \alpha\}$$

and

$$\{x \in \mathbb{R} : g(x) > \alpha\} = \bigcup_{j=1}^{\infty} \{x \in \mathbb{R} : f_j(x) > \alpha\}.$$

Thus f and g are measurable when the f_j are.

Next observe that

$$\underline{f}(x) = \sup_{j \geq 1} \left\{ \inf_{m \geq j} f_m(x) \right\}$$

and

$$\overline{g}(x) = \inf_{j \geq 1} \left\{ \sup_{m \geq j} f_m(x) \right\}.$$

This implies that $\underline{f}$ and $\overline{g}$ are measurable. $\qquad\square$

Corollary 1.21 *If $\{f_j\}$ is a sequence of $\mathcal{X}$-measurable functions on $\mathbb{R}$ which converges to a function f on $\mathbb{R}$, then f is $\mathcal{X}$-measurable.*

Proof: We note that

$$f(x) = \lim_{j \to \infty} f_j(x) = \liminf_{j \to \infty} f_j(x) = \limsup_{j \to \infty} f_j(x)$$

and the result follows. $\qquad\square$

Remark 1.22 Let us say a few words now about products of measurable functions. So assume that f and g are $\mathcal{X}$-measurable. For $j \in \mathbb{N}$, let f_j be the truncation of f defined as follows:

$$f_j(x) = \begin{cases} f(x) & \text{if} & |f(x)| \leq j, \\ j & \text{if} & f(x) > j, \\ -j & \text{if} & f(x) < -j. \end{cases}$$

Define g_j similarly.

It is easy to check that f_j and g_j are measurable (recall that characteristic functions are measurable and products and sums preserve measurability). It follows from Lemma 1.14 that $f_k \cdot g_j$ is measurable. Since

$$f(x) \cdot g_j(x) = \lim_{k \to \infty} f_k(x) \cdot g_j(x) \quad \text{for } x \in \mathbb{R},$$

it follows from Corollary 1.21 that $f \cdot g_j$ is measurable. Reasoning similarly,

$$(f \cdot g)(x) = f(x) \cdot g(x) = \lim_{j \to \infty} f(x) \cdot g_j(x) \quad \text{for } x \in \mathbb{R}.$$

Thus Corollary 1.21 implies once again that $f \cdot g$ is measurable.

Recall that, if E is a set, then χ_E is the characteristic function of E: it takes the value 1 at points of E and it takes the value 0 otherwise.

Definition 1.23 Let E_1, E_2, ..., E_k be pairwise disjoint measurable sets. Let α_1, α_2, ..., α_k be real numbers. The function

$$s(x) = \sum_{j=1}^{k} \alpha_j \, \chi_{E_j}(x)$$

is called a *simple function*.

Proposition 1.24 *Let f be a nonnegative $\mathcal{X}$-measurable function. Then there is a sequence $\{s_k\}$ of simple functions with these properties:*

(a) $0 \leq s_k(x) \leq s_{k+1}(x)$ *for $x \in \mathbb{R}$ and $k \in \mathbb{N}$;*

(b) $f(x) = \lim_{k \to \infty} s_k(x)$ *for each $x \in \mathbb{R}$;*

Proof: Fix a natural number k. For $j = 0, 1, \ldots, k \cdot 2^k - 1$, we let

$$S_{j,k} = \{x \in \mathbb{R} : j2^{-k} \leq f(x) < (j+1)2^{-k}\}.$$

Also, if $j = k \cdot 2^k$, set $S_{j,k} = \{x \in \mathbb{R} : f(x) \geq k\}$.

Notice that, for fixed k, the sets S_{jk} are pairwise disjoint for $j = 0, 1, \ldots, k \cdot 2^k$. Also each of these sets belongs to $\mathcal{X}$; and the union of the sets is all of $\mathbb{R}$. Now set

$$s_k(x) = j \cdot 2^{-k} \quad \text{for } x \in S_{j,k} , \ j = 0, 1, \ldots, k \cdot 2^k .$$

Then certainly each s_k is measurable. And properties (a) and (b) of the proposition are now immediate.   □

Exercises

1. Show that, if $a < b$, then

$$[a, b] = \bigcap_{j=1}^{\infty} (a - 1/j, b + 1/j).$$

Therefore any σ-algebra that contains all open intervals also contains all closed intervals.

Likewise, show that

$$(a, b) = \bigcup_{j=1}^{\infty} [a + 1/j, b - 1/j].$$

Hence any σ-algebra that contains all closed intervals also contains all open intervals.

2. Let $\{A_n\}_{n=1}^{\infty}$ be a collection of subsets of $\mathbb{R}$. Let $F_0 = \emptyset$. For $n = 1, 2, \ldots$, set

$$F_n = \bigcup_{j=1}^{n} A_j \quad , \quad G_n = A_n \setminus F_{n-1} \, .$$

Show that $\{F_n\}$ is a monotone increasing sequence of sets. Show that $\{G_n\}$ is a pairwise disjoint sequence of sets. Furthermore show that

$$\bigcup_{n=1}^{\infty} F_n = \bigcup_{n=1}^{\infty} G_n = \bigcup_{n=1}^{\infty} A_n \, .$$

3. Let $\{B_n\}_{n=1}^{\infty}$ be a collection of subsets of $\mathbb{R}$. Let B consist of those points $x \in \mathbb{R}$ which belong to infinitely many of the sets B_n. Show that

$$B = \bigcap_{k=1}^{\infty} \left[\bigcup_{j=k}^{\infty} B_j \right] \, .$$

We call B the *limit superior* of the sets $\{B_n\}$. This set is denoted $\limsup_{n \to \infty} B_n$.

4. Let $\{C_n\}_{n=1}^{\infty}$ be a collection of subsets of $\mathbb{R}$. Let C consist of those points $x \in \mathbb{R}$ which belong to all but finitely many of the sets C_n. Show that

$$C = \bigcup_{k=1}^{\infty} \left[\bigcap_{j=k}^{\infty} C_j \right] \, .$$

We call C the *limit inferior* of the sets $\{C_n\}$. This set is denoted $\liminf_{n \to \infty} C_n$.

5. Give an example of a function $f : \mathbb{R} \to \mathbb{R}$ such that f is *not* Borel measurable, but so that f^2 and $|f|$ *are* Borel measurable.

6. Prove that, if $f : \mathbb{R} \to \mathbb{R}$ is Borel measurable, and if $M > 0$, then the truncated function f_M defined by

$$F_M(x) = \begin{cases} f(x) & \text{if} & |f(x)| \leq M \, , \\ M & \text{if} & f(x) > M \, , \\ -M & \text{if} & f(x) < -M \, , \end{cases}$$

is also Borel measurable.

7. Let f be a nonnegative Borel measurable function on $\mathbb{R}$ which is bounded above. Show that the sequence of functions in Proposition 1.24 converges uniformly on $\mathbb{R}$ to f.

8. Let $f : \mathbb{R} \to \mathbb{R}$ be a function. If S is any subset of $\mathbb{R}$, set

$$f^{-1}(S) = \{x \in \mathbb{R} : f(x) \in S\} \, .$$

We call $f^{-1}(S)$ the *inverse image* of S under f. Show that

$$f^{-1}(\emptyset) = \emptyset \quad \text{and} \quad f^{-1}(\mathbb{R}) = \mathbb{R}.$$

If S, T are subsets of $\mathbb{R}$, then show that

$$f^{-1}(S \setminus T) = f^{-1}(S) \setminus f^{-1}(T).$$

If $\{F_j\}$ are subsets of $\mathbb{R}$, then show that

$$f^{-1}\left(\bigcup_j F_j\right) = \bigcup_j f^{-1}(F_j) \quad \text{and} \quad f^{-1}\left(\bigcap_j F_j\right) = \bigcap_j f^{-1}(F_j).$$

It follows that, if $\mathcal{Y}$ is a σ-algebra of sets in $\mathbb{R}$, then $\{f^{-1}(E) : E \in \mathcal{Y}\}$ is a σ-algebra of subsets of $\mathbb{R}$.

9. Let $\mathcal{X}$ be a σ-algebra on $\mathbb{R}$. Show that a function $f : \mathbb{R} \to \mathbb{R}$ is $\mathcal{X}$-measurable if and only if the inverse image of any Borel set lies in $\mathcal{X}$.

10. A collection $\mathcal{M}$ of subsets of $\mathbb{R}$ is called a *monotone class* if, for each monotone increasing sequence of sets $\{F_n\}_{n=1}^{\infty}$ in $\mathcal{M}$, and each monotone decreasing sequence of sets $\{G_n\}_{n=1}^{\infty}$ in $\mathcal{M}$, the sets

$$\bigcup_{n=1}^{\infty} F_n \quad \text{and} \quad \bigcap_{n=1}^{\infty} G_n$$

belong to $\mathcal{M}$. Prove that any σ-algebra is a monotone class. If now $\mathcal{C}$ is *any* collection of subsets of $\mathbb{R}$, then show that there is a smallest monotone class containing $\mathcal{C}$.

11. Fill in the details of part II of Example 1.9.

2

The Purpose of Measures

2.1   What Is a Measure?

In what follows we let $\mathbb{R}^+ = \{x \in \mathbb{R} : x \geq 0\}$. We let $\widehat{\mathbb{R}}$ denote the extended reals, and we let $\widehat{\mathbb{R}}^+ = \{x \in \widehat{\mathbb{R}} : x \geq 0\}$. Thus $\widehat{\mathbb{R}}^+ = \mathbb{R}^+ \cup \{+\infty\}$.

Definition 2.1 Let $\mathcal{X}$ be a σ-algebra on $\mathbb{R}$. A *measure* μ is a function $\mu : \mathcal{X} \to \widehat{\mathbb{R}}+$ such that

(a) $\mu(\emptyset) = 0$;

(b) If $E_1, E_2, \ldots$ are pairwise disjoint sets in $\mathcal{X}$, then

$$\mu\left(\bigcup_{j=1}^{\infty} E_j\right) = \sum_{j=1}^{\infty} \mu(E_j). \tag{2.1.1}$$

Notice that we may obtain the value $+\infty$ in equation (2.1.1) only if either (i) one of the $\mu(E_j)$ equals $+\infty$ or (ii) the sum of the $\mu(E_j)$ is $+\infty$. If a given measure μ never takes on the value $+\infty$, then we call that measure *finite*. If instead $\mathbb{R} = \cup_j E_j$ and each $\mu(E_j)$ is finite, then we say that μ is σ-finite.

EXAMPLE 2.2 Let us consider some examples of measures.

(a) Let $\mathcal{B}$ be the Borel σ-algebra on $\mathbb{R}$. Set $\mu(E) = 0$ for all Borel sets E. Then this μ is a measure (although not a very interesting one).

(b) Let $\mathcal{B}$ be the Borel σ-algebra on $\mathbb{R}$. Define $\mu(\emptyset) = 0$ and $\mu(E) = +\infty$ for every other Borel set E. Then μ is a measure. It is *not* σ-finite.

(c) Let $\mathcal{B}$ be the Borel σ-algebra on $\mathbb{R}$. Fix a point $P \in \mathbb{R}$. For E a Borel set define

$$\mu_P(E) = \begin{cases} 0 & \text{if} \quad P \notin E, \\ 1 & \text{if} \quad P \in E. \end{cases}$$

Then μ_P is a finite measure. We sometimes call this the *unit measure* or *point mass* or *Dirac mass* concentrated at P.

(d) One of the main points of this book is to construct a measure μ on the Borel σ-algebra $\mathbb{R}$ which assigns to each interval $[a, b]$ or (a, b) or $[a, b)$ or $(a, b]$ the measure $b - a$. This will be the famous *Lebesgue measure* constructed by H. Lebesgue in 1902. This is not a finite measure, but it is σ-finite because the length of each interval $I_j = [j, j + 1]$ will be 1 and $\cup_j I_j = \mathbb{R}$. We will be developing Lebesgue measure in the remainder of this book.

(e) Let g be a strictly monotone increasing function from $\mathbb{R}$ to $\mathbb{R}$. We will see later that there exists a measure μ that assigns to each interval $[a, b]$ or (a, b) or $[a, b)$ or $(a, b]$ the measure $g(b) - g(a)$. This is the *Borel-Stieltjes measure μ_g induced by g.*

(f) Let $A = \{a_j\}_{j=1}^{\infty}$ be a countable set of real numbers. For a set E in the Borel σ-algebra $\mathcal{B}$, we define $\mu(E)$ to be the number of elements of A that lies in E. Then μ is a measure.

Now we prove some basic results about measures.

Lemma 2.3 *Let $\mathcal{X}$ be a σ-algebra on $\mathbb{R}$. Let μ be a measure on $\mathcal{X}$. If $E, F \in \mathcal{X}$ and $E \subseteq F$, then $\mu(E) \leq \mu(F)$. Also, if $\mu(E) < \infty$, then $\mu(F \setminus E) = \mu(F) - \mu(E)$.*

Proof: Write $F = E \cup (F \setminus E)$ and note that $E \cap (F \setminus E) = \emptyset$. It follows that

$$\mu(F) = \mu(E) + \mu(F \setminus E). \tag{2.3.1}$$

Since $\mu(F \setminus E) \geq 0$, we conclude that $\mu(F) \geq \mu(E)$. If $\mu(E) < \infty$, then we can subtract it from both sides of equation (2.3.1) to obtain the second assertion. $\square$

Lemma 2.4 *Let μ be a measure defined on a σ-algebra $\mathcal{X}$ on $\mathbb{R}$.*

(a) *If $\{E_j\}_{j=1}^{\infty}$ is an increasing sequence of sets (i.e., $E_1 \subseteq E_2 \subseteq \cdots$) in $\mathcal{X}$, then*

$$\mu \left(\bigcup_{j=1}^{\infty} E_j \right) = \lim_{n \to \infty} \mu(E_n). \tag{2.4.1}$$

(b) *If $\{F_j\}_{j=1}^{\infty}$ is a decreasing sequence of sets (i.e., $F_1 \supseteq F_2 \supseteq \cdots$) in $\mathcal{X}$ and if $\mu(F_1) < +\infty$, then*

$$\mu \left(\bigcap_{j=1}^{\infty} F_j \right) = \lim_{n \to \infty} \mu(F_n). \tag{2.4.2}$$

Proof: For part **(a)**, if $\mu(E_j) = +\infty$ for some j, then both sides of equation (2.4.1) are $+\infty$. Thus we may assume that $\mu(E_j) < +\infty$ for all j.

Let $A_1 = E_1$ and set $A_j = E_j \setminus E_{j-1}$ for $j > 1$. Then $\{A_j\}$ is a sequence of pairwise disjoint sets in $\mathbb{R}$ so that

$$E_j = \bigcup_{m=1}^{j} A_m \quad \text{and} \quad \bigcup_{j=1}^{\infty} E_j = \bigcup_{j=1}^{\infty} A_j \,.$$

Because μ is countably additive, we see that

$$\mu\left(\bigcup_{j=1}^{\infty} E_j\right) = \mu\left(\bigcup_{j=1}^{\infty} A_j\right) = \sum_{j=1}^{\infty} \mu(A_j) = \lim_{n \to \infty} \sum_{j=1}^{n} \mu(A_j). \qquad (2.4.3)$$

By Lemma 2.3, $\mu(A_j) = \mu(E_j) - \mu(E_{j-1})$ for $j > 1$. Hence the finite series on the righthand side of equation (2.4.3) telescopes and

$$\sum_{j=1}^{n} \mu(A_j) = \mu(E_n)\,.$$

This proves equation (2.4.1).

For part **(b)**, set $E_j = F_1 \setminus F_j$, so that $\{E_j\}$ is an increasing sequence of sets in $\mathcal{X}$. We may apply part **(a)** and Lemma 2.3 to conclude that

$$\mu\left(\bigcup_{j=1}^{\infty} E_j\right) = \lim_{n \to \infty} \mu(E_n)$$

$$= \lim_{n \to \infty} [\mu(F_1) - \mu(F_n)]$$

$$= \mu(F_1) - \lim_{n \to \infty} \mu(F_n)\,. \qquad (2.4.4)$$

Since

$$\bigcup_{j=1}^{\infty} E_j = F_1 \setminus \bigcap_{j=1}^{\infty} F_j \,,$$

we may conclude that

$$\mu\left(\bigcup_{j=1}^{\infty} E_j\right) = \mu(F_1) - \mu\left(\bigcap_{j=1}^{\infty} F_j\right)\,. \qquad (2.4.5)$$

Combining (2.4.4) and (2.4.5) gives (2.4.2). $\qquad \square$

Remark 2.5 Part **(b)** of Lemma 2.4 is false without the hypothesis that $\mu(F_1) < +\infty$. For consider the example of μ being Lebesgue measure and $F_1 = [1, +\infty)$, $F_2 = [2, +\infty)$, ..., $F_j = [j, +\infty)$,.... Then notice that this is indeed a decreasing collection of sets, but $\mu(F_1) = +\infty$. And the lefthand side of (2.4.2) is $\mu(\emptyset) = 0$ while the righthand side is $\lim_{n\to\infty} \mu(F_n) = +\infty$.

Definition 2.6 A *measure space* is a triple $(\mathbb{R}, \mathcal{X}, \mu)$, where $\mathbb{R}$ is the real numbers, $\mathcal{X}$ is a σ-algebra, and μ is a measure.

And now an important and central piece of terminology.

Definition 2.7 We shall say that a certain property (P) holds μ-*almost everywhere* if there is a subset $N \subseteq \mathbb{R}$ with $\mu(N) = 0$ and so that (P) holds on $\mathbb{R} \setminus N$. For instance, we say that two functions f and g are equal μ-almost everywhere precisely when $f(x) = g(x)$ when $x \notin N$ and N has measure 0. In these circumstances we shall often write

$$f = g \quad \mu\text{-a.e.}$$

or sometimes

$$f = g \quad \text{a.e.}$$

when the measure is understood from context.

In a similar manner, we shall say that a sequence of functions $\{f_j\}$ on $\mathbb{R}$ converges μ-almost everywhere if there is a set $N \subseteq \mathbb{R}$ with $\mu(N) = 0$ and so that $\lim_{j\to\infty} f_j(x)$ exists for all $x \in \mathbb{R} \setminus N$. Call the limit function f. In these circumstances we often write

$$f = \lim_{j\to\infty} f_j(x) \quad \mu\text{-a.e.}$$

or sometimes

$$f = \lim_{j\to\infty} f_j(x) \quad \text{a.e.}$$

when the measure is understood from context.

Exercises

1. Let $\mathcal{X}$ be a σ-algebra on $\mathbb{R}$ and let μ be a measure on $\mathcal{X}$. Fix a set $K \in \mathcal{X}$. Define $\lambda(E) = \mu(K \cap E)$. Show that λ is a measure on $\mathcal{X}$.

2. Let $\mathcal{X}$ be a σ-algebra on $\mathbb{R}$ and let $\mu_1, \mu_2, \ldots, \mu_k$ be measures on $\mathcal{X}$. Let $a_1, a_2, \ldots, a_k$ be nonnegative real numbers. Show that

$$\mu = \sum_{j=1}^{k} a_j \mu_j$$

is a measure on $\mathcal{X}$.

3. Let $\mathcal{X}$ be the σ-algebra consisting of *all* subsets of $\mathbb{R}$. Define μ on $\mathcal{X}$ by **(a)** $\mu(E) = 0$ if E is denumerable and **(b)** $\mu(E) = +\infty$ if E is uncountable. Show that μ is a measure on $\mathcal{X}$.

4. Refer to Exercise 4 in Chapter 1. Let $(\mathbb{R}, \mathcal{X}, \mu)$ be a measure space. Let $\{E_j\}$ be a sequence of sets in $\mathcal{X}$. Show that

$$\mu(\liminf_{j\to\infty} E_j) \le \liminf_{j\to\infty} \mu(E_j).$$

5. Refer to Example 2.2 part **(d)**. Let μ be Lebesgue measure. If $E = \{p\}$ is a singleton set, then show that $\mu(E) = 0$. If E is countable, then show that $\mu(E) = 0$. Show that the intervals $(a, +\infty)$, $[a, +\infty)$, $(-\infty, b)$, and $(-\infty, b]$ all have measure $+\infty$.

6. Refer to Exercise 5 for terminology. If K is a compact set in $\mathbb{R}$, then show that $\mu(K) < \infty$. Show that a nonempty open set will always have positive measure.

7. Refer to Exercise 5 for terminology. What is the Lebesgue measure of the Cantor ternary set?

8. Refer to Exercise 7. Modify the construction of the Cantor set to obtain a set which has no nontrivial interval as a subset yet still has positive Lebesgue measure.

9. Refer to Exercise 5 for terminology. What is the Lebesgue measure of the set of irrational numbers?

10. Refer to Example 2.2 part **(c)**. Now define

$$\mu = \sum_{j=1}^{\infty} 2^{-j}\mu_j,$$

where μ_j is the point mass at $j \in \mathbb{Z}$. Show that μ is a measure.

11. Let $f : \mathbb{R} \to \mathbb{R}$ be a continuous function. Let μ be a Borel measure on $\mathbb{R}$. Define

$$\lambda(E) = \mu(f^{-1}(E))$$

for E a Borel set. Show that λ is a Borel measure.

3

The Lebesgue Integral

3.1   The Integration Theory of Lebesgue

Have a quick look back at Section 1.1 to remind yourself how the Riemann integral was constructed. You will now see that the Lebesgue integral is constructed rather differently.

Definition 3.1 Let $(X, \mathcal{X}, \mu)$ be a measure space. Let

$$s(x) = \sum_{j=1}^{k} a_j \chi_{E_j}(x)$$

be a simple function. We say that the simple function f is in the *standard representation* if the a_j are distinct and the E_j are pairwise disjoint. Without saying so explicitly, we will usually assume that our simple functions are in standard representation.

Now we define the integral of s with respect to the measure μ to be

$$\int s \, d\mu = \sum_{j=1}^{k} a_j \, \mu(E_j). \tag{3.1.1}$$

As noted previously, we adhere to the custom that $0 \cdot +\infty = 0$. So, for instance, the integral of the identically 0 function equals 0. It is certainly possible for the value of the integral in formula (3.1.1) to take the value $+\infty$—for instance if $a_1 = 1$ and $\mu(E_1) = +\infty$.

EXAMPLE 3.2 Let μ be Lebesgue measure on $\mathbb{R}$. Let $f : \mathbb{R} \to \mathbb{R}$ be given by

$$f(x) = \begin{cases} 2 & \text{if} & -1 < x < 1 \\ 3 & \text{if} & 3 < x < 7 \\ -1 & \text{if} & -4 \leq x < -3 \\ 0 & & \text{otherwise}. \end{cases}$$

Then

$$\int f \, d\mu(x) = 2 \cdot (1 - (-1)) + 3 \cdot (7 - 3) + (-1) \cdot ((-3) - (-4)) = 15.$$

Now we develop some elementary properties of the integral.

Lemma 3.3 Let $(X, \mathcal{X}, \mu)$ be a measure space. If φ, ψ are simple, nonnegative functions and if $c \geq 0$, then

$$\int c\varphi \, d\mu = c \int \varphi \, d\mu \qquad (3.3.1)$$

and

$$\int (\varphi + \psi) \, d\mu = \int \varphi \, d\mu + \int \psi \, d\mu \,. \qquad (3.3.2)$$

Furthermore, if λ is defined for $E \in \mathcal{X}$ by

$$\lambda(E) = \int \varphi \cdot \chi_E \, d\mu \,, \qquad (3.3.3)$$

then λ is a measure on $\mathcal{X}$.

Proof: If $c = 0$ then equation (3.3.1) becomes trivial. So suppose $c > 0$. Then $c\varphi$ is a nonnegative, simple function. If

$$\varphi(x) = \sum_{j=1}^{k} a_j \cdot \chi_{E_j}(x) \,,$$

then

$$c\varphi(x) = \sum_{j=1}^{k} ca_j \cdot \chi_{E_j}(x) \,.$$

Thus

$$\int c\varphi \, d\mu = \sum_{j=1}^{k} ca_j \cdot \mu(E_j) = c \sum_{j=1}^{k} a_j \cdot \mu(E_j) = c \int \varphi \, d\mu \,.$$

That establishes (3.3.1).

Suppose now that

$$\varphi = \sum_{j=1}^{k} a_j \cdot \chi_{E_j} \quad \text{and} \quad \psi = \sum_{\ell=1}^{m} b_\ell \cdot \chi_{F_\ell}$$

with the E_j pairwise disjoint and the F_j pairwise disjoint. Then $\varphi + \psi$ has the representation

$$\varphi + \psi = \sum_{j=1}^{k} \sum_{\ell=1}^{m} (a_j + b_\ell) \cdot \chi_{E_j \cap F_\ell} + \sum_{j=1}^{k} \sum_{\ell=1}^{m} a_j \cdot \chi_{E_j \setminus (F_1 \cup \cdots \cup F_m)}$$

$$+ \sum_{j=1}^{k} \sum_{\ell=1}^{m} b_\ell \cdot \chi_{F_\ell \setminus (E_1 \cup \cdots \cup E_k)} \,. \qquad (3.3.4)$$

This last representation for $\varphi + \psi$ is a bit confusing since different occurrences of $a_j + b_\ell$ may be equal (so that (3.3.4) is not necessarily the standard representation of simple function given in the original definition). Let c_p, $p = 1, 2, \ldots, r$, be the distinct numbers in the collection $\{a_j + b_\ell : j = $

$1, \ldots, k; \ell = 1, \ldots, m\}$. Let H_p be the union of all those sets $E_j \cap F_\ell \neq \emptyset$ so that $a_j + b_\ell = c_p$. Then

$$\mu(H_p) = \sum_{(p)} \mu(E_j \cap F_\ell).$$

Here the notation (p) means that we sum over choices of j and ℓ so that $a_j + b_\ell = c_p$ and $E_j \cap F_\ell \neq \emptyset$. Of course the standard representation of $\varphi + \psi$ is now given by

$$\varphi + \psi = \sum_{p=1}^{r} c_p \cdot \chi_{H_p} + \sum_{j=1}^{k} a_j \cdot \chi_{E_j \setminus (F_1 \cup F_2 \cup \cdots \cup F_m)} + \sum_{\ell=1}^{m} b_\ell \cdot \chi_{F_\ell \setminus (E_1 \cup E_2 \cup \cdots \cup E_k)} \cdot$$

As a result, we may now calculate that

$$\int (\varphi + \psi) \, d\mu = \sum_{p=1}^{r} c_p \cdot \mu(H_p) + \sum_{j=1}^{k} a_j \cdot \mu[E_j \setminus (F_1 \cup F_2 \cup \cdots \cup F_m)]$$

$$+ \sum_{\ell=1}^{m} b_\ell \cdot \mu[F_\ell \setminus (E_1 \cup E_2 \cup \cdots \cup E_k)]$$

$$= \sum_{p=1}^{r} \sum_{(p)} c_p \cdot \mu(E_j \cap F_\ell) + \sum_{j=1}^{k} a_j \cdot \mu[E_j \setminus (F_1 \cup F_2 \cup \cdots \cup F_m)]$$

$$+ \sum_{\ell=1}^{m} b_\ell \cdot \mu[F_\ell \setminus (E_1 \cup E_2 \cup \cdots \cup E_k)]$$

$$= \sum_{p=1}^{r} \sum_{(p)} (a_j + b_\ell) \cdot \mu(E_j \cap F_\ell) + \sum_{j=1}^{k} a_j \cdot \mu[E_j \setminus (F_1 \cup F_2 \cup \cdots \cup F_m)]$$

$$+ \sum_{\ell=1}^{m} b_\ell \cdot \mu[F_\ell \setminus (E_1 \cup E_2 \cup \cdots \cup E_k)]$$

$$= \sum_{j=1}^{k} \sum_{\ell=1}^{m} (a_j + b_\ell) \cdot \mu(E_j \cap F_\ell) + \sum_{j=1}^{k} a_j \cdot \mu[E_j \setminus (F_1 \cup F_2 \cup \cdots \cup F_m)]$$

$$+ \sum_{\ell=1}^{m} b_\ell \cdot \mu[F_\ell \setminus (E_1 \cup E_2 \cup \cdots \cup E_k)]$$

$$= \sum_{j=1}^{k} a_j \cdot \left[\mu(E_j \setminus (F_1 \cup \cdots \cup F_m)) + \sum_{\ell=1}^{m} \mu(E_j \cap F_\ell) \right]$$

$$+ \sum_{\ell=1}^{m} b_\ell \cdot \left[\mu(F_\ell \setminus (E_1 \cup \cdots \cup E_k)) + \sum_{j=1}^{k} \mu(E_j \cap F_\ell) \right]$$

$$= \sum_{j=1}^{k} a_j \cdot \mu(E_j) + \sum_{\ell=1}^{m} b_\ell \cdot \mu(F_\ell)$$

$$= \int \varphi \, d\mu + \int \psi \, d\mu.$$

To prove (3.3.3), we note that

$$\varphi \chi_E = \sum_{j=1}^{k} a_j \cdot \chi_{E_j \cap E} \, .$$

As a result, we may apply what was proved above to see that

$$\lambda(E) = \int \varphi \chi_E \, d\mu = \sum_{j=1}^{k} a_j \cdot \int \chi_{E_j \cap E} \, d\mu = \sum_{j=1}^{k} a_j \cdot \mu(E_j \cap E) \, .$$

Since the mapping $E \mapsto \mu(E_j \cap E)$ is a measure (see Exercise 1 in Chapter 2), we see that we have expressed λ as a linear combination of measures. Thus (see Exercise 2 in Chapter 2) λ is a measure. $\qquad \square$

Now we can define the integral of a nonnegative, measurable function f. The value of this integral could be finite or it could be $+\infty$.

Definition 3.4 Let $(X, \mathcal{X}, \mu)$ be a measure space and f a nonnegative, measurable function. Then the *integral of f with respect to μ* is the extended real number

$$\int f \, d\mu \equiv \sup \int s \, d\mu \, ,$$

where the supremum is taken over all nonnegative, simple functions s satisfying $0 \le s(x) \le f(x)$ for all $x \in \mathbb{R}$.

Definition 3.5 Let $(X, \mathcal{X}, \mu)$ be a measure space and f a nonnegative, measurable function. Let $E \in \mathcal{X}$. We define the *integral of f over E with respect to μ* to be

$$\int_E f \, d\mu = \int f \cdot \chi_E \, d\mu \, .$$

This is the *Lebesgue integral of f*.

Based on our experience with the Riemann integral, there are certain monotonicity properties that we expect an integral to have:

Proposition 3.6 Let $(X, \mathcal{X}, \mu)$ be a measure space.
If f and g are nonnegative, measurable functions and $f \le g$, then

$$\int f \, d\mu \le \int g \, d\mu \, . \tag{3.6.1}$$

If instead f is a nonnegative, measurable function and $E, F \in \mathcal{X}$, and if $E \subseteq F$, then

$$\int_E f \, d\mu \le \int_F f \, d\mu \, . \tag{3.6.2}$$

Proof: If s is a simple, nonnegative function such that $0 \leq s \leq f$, then it certainly follows that $0 \leq s \leq g$. Thus (3.6.1) holds.

Since $f \cdot \chi_E \leq f \cdot \chi_F$, the second assertion follows from the first. □

There are three significant theorems about the convergence of the Lebesgue integral. Contrast this situation with that for the Riemann integral—where there is really only one convergence theorem. See Section 1.1. Now we treat the first of these.

Theorem 3.7 (Lebesgue Monotone Convergence Theorem) *Let $(X, \mathcal{X}, \mu)$ be a measure space. Let $\{f_j\}$ be a monotone increasing sequence of nonnegative, measurable functions (i.e., $f_1(x) \leq f_2(x) \leq \cdots$ for all x) that converge pointwise to a function f. Then*

$$\int f \, d\mu = \lim_{j \to \infty} \int f_j \, d\mu. \tag{3.7.1}$$

Proof: According to Corollary 1.21, the function f is measurable. Since $f_j \leq f_{j+1} \leq f$, we see from Proposition 3.6 that

$$\int f_j \, d\mu \leq \int f_{j+1} \, d\mu \leq \int f \, d\mu$$

for all $j \in \mathbb{N}$. Thus we have

$$\lim_{j \to \infty} \int f_j \, d\mu \leq \int f \, d\mu. \tag{3.7.2}$$

This is half of the result. For the opposite inequality, let α be a real number satisfying $0 < \alpha < 1$ and let s be a simple function satisfying $0 \leq s \leq f$. Let

$$A_j = \{x \in X : f_j(x) \geq \alpha s(x)\}.$$

Thus $A_j \subseteq X$, $A_j \subseteq A_{j+1}$ for each j, and $X = \cup_j A_j$.

By Proposition 3.6,

$$\int_{A_j} \alpha s \, d\mu \leq \int_{A_j} f_j \, d\mu \leq \int_X f_j \, d\mu. \tag{3.7.3}$$

Since the sequence of sets $\{A_j\}$ is monotone increasing and has union X, we see from Lemma 2.4 **(a)** and (3.3.3) that

$$\int s \, d\mu = \lim_{j \to \infty} \int_{A_j} s \, d\mu.$$

Thus, taking the limit in (3.7.3) with respect to j, we find that

$$\alpha \int s \, d\mu \leq \lim_{j \to \infty} \int f_j \, d\mu.$$

Since this inequality holds for all α with $0 < \alpha < 1$, we conclude that

$$\int s \, d\mu \leq \lim_{j \to \infty} \int f_j \, d\mu .$$

Because s is an arbitrary simple function satisfying $0 \leq s \leq f$, we find now that

$$\int f \, d\mu = \sup_s \int s \, d\mu \leq \lim_{j \to \infty} \int f_j \, d\mu .$$

Combining this with (3.7.2), we obtain (3.7.1). $\qquad\qquad\qquad\square$

Remark 3.8 It may be noted that we are not assuming nor asserting that either side of (3.7.1) is finite.

EXAMPLE 3.9 Let $(X, \mathcal{X}, \mu)$ be a measure space and let μ be Lebesgue measure. Let f be a nonnegative, measurable function such that $\int f \, d\mu$ is finite. Define $f_j = f \cdot \chi_{[0,j]}$. Obviously $\lim_{j \to \infty} f_j(x) = f(x)$ for every x and the convergence is monotone increasing.

Then the lefthand side of (3.7.1) is $\int f \, d\mu$. And the righthand side is $\lim_{j \to \infty} \int f_j \, d\mu$. We are guaranteed that the limit in this righthand side equals $\int f \, d\mu$.

Our second convergence result is particularly useful because it applies to sequences of functions that are not monotone.

Theorem 3.10 (Fatou's Lemma) *Let $(X, \mathcal{X}, \mu)$ be a measure space. Assume that the functions f_j are nonnegative and measurable. Then*

$$\int (\liminf_{j \to \infty} f_j) \, d\mu \leq \liminf_{j \to \infty} \int f_j \, d\mu . \qquad (3.10.1)$$

Proof: Let $g_j(x) = \inf\{f_j(x), f_{j+1}(x), \dots\}$ for each $x \in X$. Then $g_j \leq f_k$ whenever $j \leq k$. Thus

$$\int g_j \, d\mu \leq \int f_k \, d\mu \quad \text{whenever } j \leq k .$$

Hence

$$\int g_j \, d\mu \leq \liminf_{k \to \infty} \int f_k \, d\mu .$$

Since the sequence $\{g_j\}$ is increasing and converges to $\liminf_{k \to \infty} f_k$, the monotone convergence theorem tells us that

$$\int (\liminf_{k \to \infty} f_k) \, d\mu = \lim_{j \to \infty} \int g_j \, d\mu$$

$$\leq \liminf_{k \to \infty} \int f_k \, d\mu . \qquad\qquad\square$$

Remark 3.11 Fatou's lemma is remarkable in part because it says that a liminf is greater than or equal to something else. That is not something that one often sees.

EXAMPLE 3.12 Let $\mathcal{B}$ be the Borel sets in $\mathbb{R}$ and let $d\mu$ be Lebesgue measure. Let $f_j(x) = \chi_{[j,j+1]}$ for $j = 1, 2, \dots$. Then the lefthand side of (3.10.1) is $\int 0 \, d\mu = 0$. And the righthand side of (3.10.1) is $\liminf 1 = 1$. Certainly $0 \leq 1$.

Corollary 3.13 *If f is a nonnegative, measurable function and if λ is defined on $\mathcal{X}$ by*

$$\lambda(E) = \int_E f \, d\mu = \int f \cdot \chi_E \, d\mu,$$

then λ is a measure.

Proof: Since $f \geq 0$ it is immediate that $\lambda(E) \geq 0$. If $E = \emptyset$, then $f\chi_E$ vanishes everywhere so that $\lambda(\emptyset) = 0$. To see that λ is countably additive, let $\{E_j\}$ be a pairwise disjoint sequence of sets in $\mathcal{X}$ with union E and let f_n be defined to be

$$f_n = \sum_{j=1}^{n} f \cdot \chi_{E_j}.$$

Then we see from the additivity of the integral that

$$\int f_n \, d\mu = \sum_{j=1}^{n} \int f \cdot \chi_{E_j} \, d\mu = \sum_{j=1}^{n} \lambda(E_j).$$

Since $\{f_j\}$ is an increasing sequence of nonnegative, measurable functions converging to $f\chi_E$, the monotone convergence theorem tells us that

$$\lambda(E) = \int f\chi_E \, d\mu = \lim_{j \to \infty} \int f_j \, d\mu = \lim_{j \to \infty} \sum_{\ell=1}^{j} \int f \cdot \chi_{E_\ell} \, d\mu$$

$$= \lim_{j \to \infty} \sum_{\ell=1}^{j} \lambda(E_\ell) = \sum_{\ell=1}^{\infty} \lambda(E_\ell). \qquad \square$$

Corollary 3.14 *Let $(X, \mathcal{X}, \mu)$ be a measure space. Suppose that f is a nonnegative, measurable function. Then $f(x) = 0$ μ-almost everywhere if and only if*

$$\int f \, d\mu = 0. \tag{3.14.1}$$

Proof: If equation (3.14.1) holds, then we let

$$E_j = \left\{ x \in X : f(x) > \frac{1}{j} \right\}.$$

As a result, $f \geq (1/j)\chi_{E_j}$ for each j.

Therefore

$$0 = \int f \, d\mu \geq \int_{E_j} f \, d\mu \geq \int_{E_j} \frac{1}{j} \, d\mu = \frac{1}{j} \cdot \mu(E_j) \geq 0 .$$

We conclude that $\mu(E_j) = 0$, hence the set

$$\{x \in X : f(x) > 0\} = \bigcup_{j=1}^{\infty} E_j$$

has measure 0.

For the converse, assume that $f(x) = 0$ almost everywhere. If

$$E = \{x \in X : f(x) > 0\} ,$$

then $\mu(E) = 0$. Set $f_j = j\chi_E$ for $j = 1, 2, \ldots$. Since $f \leq \liminf_{j \to \infty} f_j$, Fatou's lemma implies that

$$0 \leq \int f \, d\mu \leq \int \liminf_{j \to \infty} f_j \, d\mu \leq \liminf_{j \to \infty} \int f_j \, d\mu = 0 .$$

Thus $\int f \, d\mu = 0$.   □

Corollary 3.15 *Let $(X, \mathcal{X}, \mu)$ be a measure space. Let g_j be a sequence of nonnegative, measurable functions. Then*

$$\int \left(\sum_{j=1}^{\infty} g_j \right) d\mu = \sum_{j=1}^{\infty} \left(\int g_j \, d\mu \right) .$$

Proof: Set $f_j = g_1 + g_2 + \cdots + g_j$. Now apply the monotone convergence theorem to the f_j.   □

Exercises

1. Show that the collection of simple functions is closed under the operations of sum, scalar multiplication, and product.

2. Suppose that f and g are simple functions. Then show that

$$F = \max\{f, g\} \quad \text{and} \quad G = \min\{f, g\}$$

are both simple functions.

3. Let $(\mathbb{R}, \mathcal{X}, \mu)$ be a measure space and assume that μ is Lebesgue measure. Let $a_j \geq 0$ and assume that $\sum_j a_j$ converges. Define

$$f(x) = a_j \quad \text{if} \quad j \leq x < j+1$$

for $j = 1, 2, \ldots$. Show that

$$\int f \, d\mu = \sum_{j=1}^{\infty} a_j.$$

4. Let $(\mathbb{R}, \mathcal{X}, \mu)$ be a measure space and assume that μ is Lebesgue measure. Set $f_j = \chi_{[0,j]}$. This sequence of functions is monotone increasing to $f = \chi_{[0,\infty]}$. Notice that $\int f_j \, d\mu < \infty$ for each j but $\int f \, d\mu = +\infty$. Does the monotone convergence theorem apply here? Why or why not?

5. Let $(\mathbb{R}, \mathcal{X}, \mu)$ be a measure space and assume that μ is Lebesgue measure. Set $f_j = (1/j)\chi_{[j,+\infty)}$. Then the sequence $\{f_j\}$ is monotone *decreasing*. And it converges uniformly to the identically 0 function $f \equiv 0$. But

$$0 = \int f \, d\mu \neq \lim_{j \to \infty} \int f_j \, d\mu = +\infty.$$

We conclude that there is no monotone convergence theorem for decreasing sequences of functions. Provide the details.

6. Let $(\mathbb{R}, \mathcal{X}, \mu)$ be a measure space and assume that μ is Lebesgue measure. Set $f_j = (1/j)\chi_{[j,+\infty]}$ and $f \equiv 0$. Prove that $f_j \to f$ uniformly, but

$$\int f \, d\mu \neq \lim_{j \to \infty} \int f_j \, d\mu.$$

Why does this result not contradict the monotone convergence theorem? Is Fatou's lemma relevant?

7. Notation is as in Exercise 6. Set $g_j = j\chi_{[1/j,2/j]}$. Also set $g \equiv 0$. Prove that

$$\int g \, d\mu \neq \lim_{j \to \infty} \int g_j \, d\mu.$$

Does the sequence $\{g_j\}$ converge uniformly to g? Is the monotone convergence theorem relevant? Does Fatou's lemma apply?

8. Assume that $(X, \mathcal{X}, \mu)$ is a finite measure space. Let $\{f_j\}$ be a sequence of nonnegative, measurable functions which converges uniformly to a limit function f. Prove then that f is a nonnegative, measurable function and that

$$\int f \, d\mu = \lim_{j \to \infty} \int f_j \, d\mu.$$

9. Let X be the closed interval $[a, b]$ of finite length. Let $\mathcal{B}$ be the Borel sets in X and let μ be Lebesgue measure. Suppose that f is a nonnegative, continuous function on X. Prove that

$$\int_X f \, d\mu = \int_a^b f(x) \, dx,$$

where the integral on the right is the classical Riemann integral. [**Hint:** First prove the result when f is a linear combination of characteristic functions of intervals.]

10. Let $(\mathbb{R}, \mathcal{X}, \mu)$ be a measure space and suppose that μ is Lebesgue measure. Set $f_j = (-1/j)\chi_{[0,j]}$. This sequence f_j converges uniformly to $f \equiv 0$ on $[0, \infty)$. But $\int f_j \, d\mu = -1$ whereas $\int f \, d\mu = 0$. Thus

$$\liminf_{j \to \infty} \int f_j \, d\mu = -1 < 0 = \int f \, d\mu.$$

We see then that Fatou's lemma may be false if the hypothesis $f_j \geq 0$ fails.

11. Prove that if f is a nonnegative, measurable function and

$$\int f \, d\mu < +\infty,$$

then the set

$$Z = \{x \in X : f(x) > 0\}$$

is σ-finite.

12. Let $(\mathbb{R}, \mathcal{X}, \mu)$ be a measure space and assume that μ is Lebesgue measure. Show that if f is a nonnegative, measurable function and

$$\int f \, d\mu < +\infty,$$

then, for any $\epsilon > 0$, there exists a measurable set E such that $\mu(E) < +\infty$ and

$$\int f \, d\mu \leq \int_E f \, d\mu + \epsilon.$$

4

Integrable Functions

4.1 Functions with Finite Integral

In earlier parts of the book we considered the integral of a nonnegative function. In the present chapter we shall treat functions that take both positive and negative values.

Definition 4.1 Let $(X, \mathcal{X}, \mu)$ be a measure space. Let $f : \mathbb{R} \to \mathbb{R}$ be a measurable function. We remind the reader that

$$f^+(x) = \max\{f(x), 0\} = \frac{f(x) + |f(x)|}{2}$$

and

$$f^-(x) = \max\{-f(x), 0\} = \frac{|f(x)| - f(x)}{2}.$$

Note that f^+ is the positive part of f and f^- is the negative part of f. Of course we have that $f = f^+ - f^-$.

We call f *integrable* if both f^+ and f^- have finite integral. In that case we set

$$\int f \, d\mu = \int f^+ \, d\mu - \int f^- \, d\mu.$$

If E is a measurable set then we define

$$\int_E f \, d\mu = \int_E f^+ \, d\mu - \int_E f^- \, d\mu = \int f^+ \cdot \chi_E \, d\mu - \int f^- \cdot \chi_E \, d\mu.$$

Definition 4.2 Let $\mathcal{X}$ be a σ-algebra on $\mathbb{R}$. A *signed measure* on $\mathcal{X}$ is defined to be a function $\lambda : \mathcal{X} \to \mathbb{R}$ so that

(a) $\lambda(\emptyset) = 0$;

(b) If $E_1, E_2, \ldots$ are pairwise disjoint sets in $\mathcal{X}$, then

$$\lambda \left(\bigcup_{j=1}^{\infty} E_j \right) = \sum_{j=1}^{\infty} \lambda(E_j).$$

The difference between a signed measure and a measure is that a signed measure can take *any real value* while a measure can only take nonnegative values. For technical reasons, we *do not* allow a signed measure to take the values $\pm\infty$ (while a measure *is* allowed to take the value $+\infty$). We shall usually denote a measure by μ and a signed measure by λ.

Remark 4.3 We make it an exercise for the reader to check the following. Suppose that f is as in Definition 4.1. Also assume that $f = f_1 - f_2$, with both f_1 and f_2 nonnegative functions having finite integral. Then

$$\int f\, d\mu = \int f^+ \, d\mu - \int f^- \, d\mu = \int f_1 \, d\mu - \int f_2 \, d\mu \,.$$

Lemma 4.4 *If f is integrable and $\lambda : \mathcal{X} \to \mathbb{R}$ is defined by*

$$\lambda(E) = \int_E f\, d\mu \,,$$

then λ is a signed measure.

Proof: Since f^+ and f^- are positive, measurable functions, then Corollary 3.13 tells us that

$$\lambda^+(E) \equiv \int_E f^+ \, d\mu \quad \text{and} \quad \lambda^-(E) \equiv \int_E f^- \, d\mu$$

are measures on $\mathcal{X}$. They are finite because f is integrable. Since $\lambda = \lambda^+ - \lambda^-$, it follows that λ is a signed measure. $\qquad\square$

Theorem 4.5 *A measurable function f is integrable if and only if $|f|$ is integrable. In this case,*

$$\left| \int f\, d\mu \right| \le \int |f| \, d\mu \,. \tag{4.5.1}$$

Proof: By definition, f is integrable if and only if both f^+ and f^- have finite integral. Since

$$|f|^+ = |f| = f^+ + f^-$$

and since

$$|f|^- = 0 \,,$$

we see that Proposition 3.6 and the additivity of the integral imply the asserted inequality.

That is to say,

$$
\left| \int f \, d\mu \right| = \left| \int f^+ \, d\mu - \int f^- \, d\mu \right|
$$

$$
\leq \int f^+ \, d\mu + \int f^- \, d\mu
$$

$$
= \int |f| \, d\mu \, . \qquad \square
$$

Corollary 4.6 *If f is measurable, g is integrable, and $|f| \leq g$, then f is integrable and*

$$
\int |f| \, d\mu \leq \int g \, d\mu \, .
$$

Proof: The result is immediate from Proposition 3.6. $\qquad \square$

Remark 4.7 It follows from our earlier discussions that the integral respects scalar multiplication and addition. We shall say no more about the matter at this time.

We next treat the most important and versatile convergence theorem for integrable functions.

Theorem 4.8 (Lebesgue Dominated Convergence Theorem) *Let $\{f_j\}$ be a sequence of integrable functions which converges almost everywhere to a measurable function f. If there is an integrable function g such that $|f_j| \leq g$ for all j, then f is integrable and*

$$
\int f \, d\mu = \lim_{j \to \infty} \int f_j \, d\mu \, . \qquad (4.8.1)
$$

Proof: By simply redefining f_j, f to equal 0 on a set of measure 0, we may assume that the convergence takes place on all of X. We may infer from Corollary 4.6 that f is integrable. Since $g + f_j \geq 0$ for each j, we may apply Fatou's lemma to find that

$$
\int g \, d\mu + \int f \, d\mu \;=\; \int (g + f) \, d\mu
$$

$$
=\; \int \liminf_{j \to \infty} (g + f_j) \, d\mu
$$

$$
\leq\; \liminf_{j \to \infty} \int (g + f_j) \, d\mu
$$

$$
=\; \liminf_{j \to \infty} \left(\int g \, d\mu + \int f_j \, d\mu \right)
$$

$$
=\; \int g \, d\mu + \liminf \int f_j \, d\mu \, .
$$

It follows that

$$\int f \, d\mu \le \liminf_{j \to \infty} \int f_j \, d\mu . \tag{4.8.2}$$

Since $g - f_j \ge 0$ for each j, we may again apply Fatou's lemma as well as the additivity of the integral to obtain

$$\begin{aligned}
\int g \, d\mu - \int f \, d\mu &= \int (g - f) \, d\mu \\
&= \int \liminf_{j \to \infty} (g - f_j) \, d\mu \\
&\le \liminf_{j \to \infty} \int (g - f_j) \, d\mu \\
&= \int g \, d\mu - \limsup_{j \to \infty} \int f_j \, d\mu .
\end{aligned}$$

From this we may infer that

$$\limsup_{j \to \infty} \int f_j \, d\mu \le \int f \, d\mu . \tag{4.8.3}$$

Now, combining (4.8.2) and (4.8.3), we conclude that

$$\int f \, d\mu = \lim_{j \to \infty} \int f_j \, d\mu . \qquad \square$$

EXAMPLE 4.9 We again work with Lebesgue measure on the real line. Let $f_j(x) = \chi_{[j, j+1]}$ for $j = 1, 2, \dots$. Then it is easy to see that there is no integrable function g with $|f_j| \le g$ for all j. Thus we *may not* apply the dominated convergence theorem to conclude that

$$\lim_{j \to \infty} \int f_j \, d\mu = \int \lim_{j \to \infty} f_j \, d\mu ,$$

and in fact they are not equal.

Remark 4.10 In the next two examples we shall use a version of the Lebesgue Dominated Convergence Theorem that is a bit different from the formulation in Theorem 4.8. Namely, instead of a sequence of functions $f_j(x)$ as $j \to \infty$ we shall instead have a continuum of functions f_t parametrized by a parameter $t \in \mathbb{R}$ and consider $\lim_{t \to t_0} f_t(x)$. These two processes are in fact logically equivalent just because

$$\lim_{t \to t_0} f_t(x) = \ell \quad \text{if and only if} \quad \lim_{j \to \infty} f_{t_j}(x) = \ell \text{ for each sequence } t_j \to t_0 .$$

EXAMPLE 4.11 We work with Lebesgue measure on the real line. Let us show that the function

$$F(t) = \int_{(0,\infty)} e^{-x} \cos(\pi tx) \, d\mu(x)$$

is continuous.

We intend to apply the dominated convergence theorem with

$$g(x) = \chi_{[0,\infty)}(x) \cdot e^{-x} = \begin{cases} e^{-x} & \text{for} & x \geq 0 \\ 0 & \text{for} & x < 0 \,. \end{cases}$$

Obviously g is measurable and $\chi_{[0,\infty)} \cdot e^{-x} \cdot \cos(\pi t)$ is measurable for each t because e^{-x} is continuous and $\chi_{[0,\infty)}$ is measurable. We need to know that $\int_{\mathbb{R}} g \, d\mu < \infty$. But the monotone convergence theorem tells us that

$$\int_{\mathbb{R}} g \, d\mu = \lim_{j \to \infty} \int_{\mathbb{R}} \chi_{[-j,j]} \cdot g \, d\mu = \lim_{j \to \infty} \int_{\mathbb{R}} \chi_{[0,j]} \cdot e^{-x} \, d\mu(x) = \lim_{j \to \infty} \int_0^j e^{-x} \, d\mu(x) \,.$$

The limit is of course 1 by the theory of the Riemann integral. So $\int_{\mathbb{R}} g \, d\mu < \infty$.

Now we may apply the dominated convergence theorem just because

$$|\chi_{[0,\infty)}(x) \cdot e^{-x} \cdot \cos(\pi tx)| \leq g(x)$$

for each $(x,t) \in \mathbb{R}^2$ (noting that $t \mapsto \chi_{[0,\infty)}(x) e^{-x} \cos(\pi tx)$ is continuous for each fixed x). We conclude that

$$\lim_{t \to t_0} F(t) = F(t_0) \,.$$

EXAMPLE 4.12 A very standard operation in mathematical analysis is "differentiating under the integral sign." In this example we use the dominated convergence theorem to analyze and justify this operation.

Assume that $f : \mathbb{R} \times \mathbb{R} \to \mathbb{R}$ satisfies

(a) $x \mapsto f^t(x) \equiv f(x,t)$ is measurable for each fixed $t \in \mathbb{R}$;

(b) $f^{t_0}(x) \equiv f(x,t_0)$ is integrable for some fixed $t_0 \in \mathbb{R}$;

(c) $\partial f(x,t)/\partial t$ exists for each (x,t).

Further suppose that there is an integrable function $g : \mathbb{R} \to \mathbb{R}$ so that

$$\left| \frac{\partial f}{\partial t}(x,t) \right| \leq g(x)$$

for each $x, t \in \mathbb{R}$.

Then the function $x \mapsto f(x,t)$ is integrable for each t and the function

$$F(t) = \int_{\mathbb{R}} f^t \, d\mu = \int_{\mathbb{R}} f(x,t) \, d\mu(x)$$

is differentiable with derivative

$$F'(t) = \frac{d}{dt} \int_{\mathbb{R}} f(x,t) \, d\mu(x) = \int_{\mathbb{R}} \frac{\partial}{\partial t} f(x,t) \, d\mu(x). \qquad (4.12.1)$$

It is easy to see from equation (4.12.1) why this phenomenon is called differentiation under the integral sign. Now we shall use the theory developed so far to see why this process is correct.

For each $t \ne t_0$, apply the mean value theorem to the function $t \mapsto f(x,t)$ to find a number c between t_0 and t so that

$$f(x,t) - f(x,t_0) = \left[\frac{\partial f}{\partial t}(x,c)\right] \cdot (t - t_0).$$

It follows that

$$|f(x,t) - f(x,t_0)| \le g(x) \cdot |t - t_0|$$

hence

$$|f(x,t)| \le |f(x,t_0)| + g(x) \cdot |t - t_0|.$$

We conclude that

$$\int_{\mathbb{R}} |f(x,t)| \, d\mu(x) \le \int_{\mathbb{R}} (|f(x,t_0)| + g(x) \cdot |t - t_0|) \, d\mu(x)$$

$$= \int |f(x,t_0)| \, d\mu(x) + |t - t_0| \int_{\mathbb{R}} g(x) \, d\mu(x).$$

This shows that the function $x \mapsto f(x,t)$ is integrable for each fixed t.

To establish the formula for F', consider any sequence $\{t_j\}_{j=1}^{\infty}$ with $\lim_{j\to\infty} t_j = t$ and $t_j \ne t$ for each j. We claim that

$$\lim_{j\to\infty} \frac{F(t_j) - F(t)}{t_j - t} = \int_{\mathbb{R}} \frac{\partial}{\partial t} f(x,t) \, d\mu(x). \qquad (4.12.2)$$

In fact we have

$$\frac{F(t_j) - F(t)}{t_j - t} = \int_{\mathbb{R}} \frac{f(x,t_j) - f(x,t)}{t_j - t} \, d\mu(x) \equiv \int_{\mathbb{R}} f_j(x,t) \, d\mu(x)$$

where

$$f_j(x,t) = \frac{f(x,t_j) - f(x,t)}{t_j - t}.$$

Observe that, for each fixed x, we know that

$$\lim_{j\to\infty} f_j(x,t) = \frac{\partial f}{\partial t}(x,t)$$

and hence (4.12.1) will follow from the dominated convergence theorem once we establish that $|f_j(x,t)| \le g(x)$ for each x.

But that claim follows from another application of the mean value theorem because there is a c' between t and t_j (with c' of course depending on x and t) such that

$$f_j(x,t) = \frac{f(x,t_j) - f(x,t)}{t_j - t} = \frac{\partial f}{\partial t}(x,c').$$

Thus $|f_j(x,t)| \le g(x)$ for each x.

Exercises

1. Let $(X, \mathcal{X}, \mu)$ be a measure space. Suppose that f is an integrable function. Let $\alpha > 0$. Show that the set $\{x \in \mathbb{R} : |f(x)| > \alpha\}$ has finite measure. Moreover, show that $\{x \in \mathbb{R} : f(x) \ne 0\}$ is σ-finite.

2. Let $(X, \mathcal{X}, \mu)$ be a measure space. Let f be an integrable function, and let $\epsilon > 0$. Show that there is a simple function s such that

$$\int |f - s| \, d\mu < \epsilon.$$

3. Let $(X, \mathcal{X}, \mu)$ be a measure space. Let f be an integrable function. Let g be a bounded, measurable function. Then show that $f \cdot g$ is integrable.

4. Let $(X, \mathcal{X}, \mu)$ be a measure space. Suppose that f is an integrable function. Does it follow that f^2 is integrable? What about other powers of f?

5. Let $(X, \mathcal{X}, \mu)$ be a measure space. Let f_1, f_2 be integrable functions. Assume that

$$\int_E f_1 \, d\mu = \int_E f_2 \, d\mu$$

for every set $E \in \mathcal{X}$. Then prove that $f_1 = f_2$ almost everywhere.

6. Let $(X, \mathcal{X}, \mu)$ be a measure space. Assume that $\mu(X) < \infty$. Let f_j be a sequence of integrable functions that converges uniformly on X to a limit function f. Prove that

$$\int f \, d\mu = \lim_{j \to \infty} \int f_j \, d\mu.$$

7. Show that the condition $\mu(X) < \infty$ cannot be dropped in Exercise 6.

8. Show that the condition $|f_j| \le g$ cannot be dropped in the Lebesgue dominated convergence theorem.

9. Let $(X, \mathcal{X}, \mu)$ be a measure space. Suppose that the f_j are integrable functions and that

$$\sum_{j=1}^{\infty} \int |f_j| \, d\mu < +\infty \, .$$

Prove that the series $\sum_j f_j(x)$ converges almost everywhere to an integrable function f.

10. Let $(X, \mathcal{X}, \mu)$ be a measure space. Let the functions f_j be integrable and suppose that the f_j converge pointwise to a function f. Show that if

$$\lim_{j \to \infty} \int |f_j - f| \, d\mu = 0 \, ,$$

then

$$\int |f| \, d\mu = \lim_{j \to \infty} \int |f_j| \, d\mu \, .$$

11. Prove that

$$\int_0^{+\infty} x^n e^{-x} \, dx = n! \, .$$

12. Let $(X, \mathcal{X}, \mu)$ be a measure space. Let f be a measurable function. For $j \in \mathbb{N}$, let

$$f_j(x) = \begin{cases} f(x) & \text{if} & |f(x)| \le j \, , \\ j & \text{if} & f(x) > j \, , \\ -j & \text{if} & f(x) < -j \, . \end{cases}$$

If f is integrable, then prove that

$$\int f \, d\mu = \lim_{j \to \infty} \int f_j \, d\mu \, .$$

Conversely, if

$$\sup_j \int |f_j| \, d\mu < +\infty \, ,$$

then prove that f is integrable.

5

The Lebesgue Spaces

5.1 Definition of the Spaces

In this chapter we are going to discuss infinite-dimensional vector spaces of functions. To this end we begin by reviewing the concept of vector space.

Definition 5.1 A *vector space* over the field $\mathbb{R}$ is a set V together with a binary operation of addition (denoted $+$) and a second operation of scalar multiplication (denoted $\cdot$) so that the following axioms are satisfied. Let $\mathbf{u}, \mathbf{v}, \mathbf{w}$ be elements of V (we call these *vectors*) and let a, b be scalars (i.e., real numbers). Then

(a) $\mathbf{u} + (\mathbf{v} + \mathbf{w}) = (\mathbf{u} + \mathbf{v}) + \mathbf{w};$

(b) $\mathbf{u} + \mathbf{v} = \mathbf{v} + \mathbf{u};$

(c) There exists an element $05 \in V$, called the *zero vector*, so that $\mathbf{v} + 0 = \mathbf{v}$ for any $\mathbf{v} \in V$.

(d) For every $\mathbf{v} \in V$, there is an element $-\mathbf{v} \in V$, called the *additive inverse* of $\mathbf{v}$, so that $\mathbf{v} + (-\mathbf{v}) = 0$.

(e) $a \cdot (b \cdot \mathbf{v}) = (a \cdot b) \cdot \mathbf{v};$

(f) $1 \cdot \mathbf{v} = \mathbf{v}$, where 1 denotes the number $1 \in \mathbb{R}$.

(g) $a \cdot (\mathbf{u} + \mathbf{v}) = a \cdot \mathbf{u} + a \cdot \mathbf{v};$

(h) $(a + b) \cdot \mathbf{u} = a \cdot \mathbf{u} + b \cdot \mathbf{u}.$

Often in practice we omit the $\cdot$ when writing scalar multiplication.

EXAMPLE 5.2 Of course $\mathbb{R}^N$ is a vector space over $\mathbb{R}$. The addition operation is

$$\langle a_1, a_2, \ldots, a_N \rangle + \langle b_1, b_2, \ldots, b_N \rangle = \langle a_1 + b_1, a_2 + b_2, \ldots, a_N + b_N \rangle.$$

The operation of scalar multiplication is

$$c \langle a_1, a_2, \ldots, a_N \rangle = \langle ca_1, ca_2, \ldots, ca_N \rangle.$$

The space ℓ^1 of all sequences $\{a_j\}_{j=1}^{\infty}$ such that $\sum_j |a_j| < \infty$ is a vector space. The operation of addition is

$$\{a_j\}_{j=1}^{\infty} + \{b_j\}_{j=1}^{\infty} = \{a_j + b_j\}_{j=1}^{\infty} \, .$$

The scalar multiplication operation is

$$c\{a_j\}_{j=1}^{\infty} = \{ca_j\}_{j=1}^{\infty} \, .$$

Definition 5.3 Let V be a vector space. A real-valued function N on V is said to be a *norm* if

(a) $N(\mathbf{v}) \geq 0$ for all $\mathbf{v} \in V$.

(b) $N(\mathbf{v}) = 0$ if and only if $\mathbf{v} = 0$.

(c) $N(\alpha \mathbf{v}) = |\alpha| N(\mathbf{v})$ for all $\mathbf{v} \in V$ and all real α.

(d) $N(\mathbf{u} + \mathbf{v}) \leq N(\mathbf{u}) + N(\mathbf{v})$ for all $\mathbf{u}, \mathbf{v} \in V$.

It is quite standard in many contexts to denote $N(\mathbf{v})$ by $\|\mathbf{v}\|$.

A *normed linear space* is a vector space equipped with a norm. If condition (b) for a norm is dropped then N is called a *seminorm*.

EXAMPLE 5.4 If $V = \mathbb{R}^N$ and $\mathbf{v} = \langle x_1, x_2, \dots, x_N \rangle \in V$, then we set

$$N(\mathbf{v}) = \sqrt{x_1^2 + x_2^2 + \cdots + x_N^2} \, .$$

It is common to denote this norm on $\mathbb{R}^N$ by $\|\mathbf{v}\|$.

Another norm for $\mathbb{R}^N$ is given, for $p \geq 1$, by

$$\|\mathbf{v}\|_p = (|x_1|^p + |x_2|^p + \cdots + |x_N|^p)^{1/p} \, .$$

We shall not provide the details of the triangle inequality, but leave that matter as an exercise for you to think about.

The space ℓ^1, which we described above, has the norm

$$\|\{a_j\}\|_1 = \sum_j |a_j| \, .$$

The space X of functions f on $[0, 1]$ so that f' exists and is continuous forms a vector space. The expression

$$N(f) = \sup_{x \in [0,1]} |f'(x)|$$

is a seminorm on X. For if f is a (nonzero) constant function, then

$$N(f) = 0 \, .$$

Definition 5.5 Let $(X, \mathcal{X}, \mu)$ be a measure space. Consider the space of all integrable functions. It is common to denote this space by $L^1(X, \mu)$ or just L^1. The norm on L^1 is

$$\|f\|_{L^1} = \int |f| \, d\mu \, .$$

In fact it is convenient to identify two integrable functions if they are equal almost everywhere. This is an equivalence relation on the set of all integrable functions. And we think, in practice, of L^1 as the collection of such equivalence classes.

Definition 5.6 Let $(X, \mathcal{X}, \mu)$ be a measure space. Let $1 \le p < \infty$. The space of all measurable functions f such that $|f|^p$ has finite integral is denoted by $L^p(X, \mu)$ or simply L^p. The norm on this space is

$$\|f\|_{L^p} = \int |f|^p \, d\mu^{1/p} \, .$$

We shall now prove a sequence of lemmas that will establish, among other things, the non-obvious fact that $\| \ \|_{L^p}$ is actually a norm.

Proposition 5.7 (Hölder's Inequality) *Let $f \in L^p$ and $g \in L^q$, where $1 < p < \infty$, $1 < q < \infty$, and $1/p + 1/q = 1$. Then $f \cdot g \in L^1$ and*

$$\|f \cdot g\|_{L^1} \le \|f\|_{L^p} \cdot \|g\|_{L^q} \, .$$

Proof: Let α be a real number with $0 < \alpha < 1$. Consider the function

$$\varphi(t) = \alpha t - t^\alpha$$

for $t \ge 0$. One may check that $\varphi'(t) < 0$ for $0 < t < 1$ and $\varphi'(t) > 0$ for $t > 1$. The mean value theorem then implies that $\varphi(t) \ge \varphi(1)$ and that $\varphi(t) = \varphi(1)$ if and only if $t = 1$. We conclude then that

$$t^\alpha \le \alpha t + (1 - \alpha) \quad \text{for} \ \ t \ge 0 \, .$$

Let $a \ge 0$, $b > 0$, and set $t = a/b$ in this last inequality. Multiply through by b. The result is

$$a^\alpha \cdot b^{1-\alpha} \le \alpha a + (1 - \alpha)b \, . \tag{5.7.1}$$

Note that equality holds here if and only if $a = b$.

Now let $1 < p < \infty$ and $1/p + 1/q = 1$. Set $\alpha = 1/p$. If A, B are nonnegative numbers and if we set $a = A^{1/\alpha} = A^p$ and $B = B^{1/(1-\alpha)} = B^{p/(p-1)} = B^q$, then we may conclude from (5.7.1) that

$$AB \le \frac{A^p}{p} + \frac{B^q}{q} \tag{5.7.2}$$

and that equality holds if and only if $A^p = B^q$.

Now suppose that $f \in L^p$ and $g \in L^q$ and that $\|f\|_{L^p} \neq 0$ and $\|g\|_{L^q} \neq 0$. The product of these functions is certainly measurable and (5.7.2) with $A = |f(x)|/\|f\|_{L^p}$, $B = |g(x)|/\|g\|_{L^q}$ tells us that

$$\frac{|f(x) \cdot g(x)|}{\|f\|_{L^p}\|g\|_{L^q}} \leq \frac{|f(x)|^p}{p\|f\|_{L^p}^p} + \frac{|g(x)|^q}{q\|g\|_{L^q}^q}.$$

Since both terms on the righthand side of this last inequality are integrable, it follow from Corollary 4.6 and the additivity of the integral that fg is integrable. On performing the integral we find that

$$\frac{\|fg\|_{L^1}}{\|f\|_{L^p}\|g\|_{L^q}} \leq \frac{1}{p} + \frac{1}{q} = 1.$$

This is Hölder's inequality. □

Corollary 5.8 (Cauchy-Schwarz-Bunyakovskii) *If f and g both belong to L^2, then $f \cdot g$ is integrable and*

$$\left| \int fg \, d\mu \right| \leq \int |fg| \, d\mu \leq \|f\|_{L^2} \cdot \|g\|_{L^2}.$$

It is worth noting that the theorem is trivially true when $p = 1$ and $q = \infty$ or $p = \infty$ and $q = 1$. We shall treat the space L^∞ in some detail below.
The next result shows that the L^p norms satisfy a triangle inequality.

Proposition 5.9 (Minkowski's Inequality) *If the functions f and g both belong to L^p, $p \geq 1$, then $f + g$ also belongs to L^p and*

$$\|f + g\|_{L^p} \leq \|f\|_{L^p} + \|g\|_{L^p}. \tag{5.9.1}$$

Proof: The case $p = 1$ is easy, so we concentrate on $p > 1$. The sum $f + g$ is plainly measurable. Since

$$|f + g|^p \leq [2 \max\{|f|, |g|\}]^p \leq 2^p\{|f|^p + |g|^p\},$$

it follows from Corollary 4.6 and the additivity of the integral that $f + g \in L^p$. Furthermore,

$$|f + g|^p = |f + g| \cdot |f + g|^{p-1} \leq |f| \cdot |f + g|^{p-1} + |g| \cdot |f + g|^{p-1}. \tag{5.9.2}$$

Since $f + g \in L^p$, we see that $|f + g|^p \in L^1$. Furthermore, since $p = (p-1)q$, it follows that $|f + g|^{p-1} \in L^q$. Thus we can apply Hölder's inequality to conclude that

$$\int |f||f + g|^{p-1} \, d\mu \leq \|f\|_{L^p} \cdot \left\{ \int |f + g|^{(p-1)q} \, d\mu \right\}^{1/q}$$

$$= \|f\|_{Lp} \cdot \|f + g\|_{L^p}^{p/q}.$$

If we treat the second term on the right in (5.9.2) similarly, the result is

$$\|f + g\|_{L^p}^p \leq \|f\|_{L^p} \cdot \|f + g\|_{L^p}^{p/q} + \|g\|_{L^p}\|f + g\|_{L^p}^{p/q}$$
$$= (\|f\|_{L^p} + \|g\|_{L^p}) \cdot \|f + g\|_{L^p}^{p/q}.$$

If $M = \|f + g\|_{L^p} = 0$, then equation (5.9.1) is trivial. If instead $M \neq 0$, then we can divide the last inequality by $M^{p/q}$. Since $p - p/q = 1$, Minkowski's inequality results. □

A *Banach space* is a normed linear space that is complete. This means that any Cauchy sequence has a limit *in that space*. The theory of Banach spaces is rich and fertile. It is a powerful tool in mathematical analysis. Our next task is to show that the L^p spaces are Banach spaces.

Definition 5.10 A sequence $\{f_j\} \subseteq L^p$ is said to be *convergent* to $f \in L^p$ if, for every $\epsilon > 0$, there is a number $J > 0$ so that if $j > J$, then $\|f_j - f\|_{L^p} < \epsilon$.

The sequence $\{f_j\}$ in L^p is a *Cauchy sequence* if, for every $\epsilon > 0$, there is a number $J > 0$ so that if $j, k > J$, then $\|f_j - f_k\|_{L^p} < \epsilon$.

Definition 5.11 The space L^p is *complete* if every Cauchy sequence in L^p converges to an element $f \in L^p$.

Lemma 5.12 *If the sequence $\{f_j\} \subseteq L^p$ converges to $f \in L^p$, then the sequence is Cauchy.*

Proof: Let $\epsilon > 0$. Choose $J > 0$ so that $j > J$ implies that $\|f_j - f\|_{L^p} < \epsilon/2$. Now let $j, k > J$. Then

$$\|f_j - f_k\|_{L^p} \leq \|f_j - f\|_{L^p} + \|f - f_k\|_{L^p} < \frac{\epsilon}{2} + \frac{\epsilon}{2} = \epsilon.$$

It follows that the sequence $\{f_j\}$ is Cauchy. □

Theorem 5.13 *Let $1 \leq p < \infty$. Then the space L^p is a complete, normed, linear space.*

Proof: Let $\{f_j\}$ be a Cauchy sequence in the L^p norm. Our job is to show that there exists an $f \in L^p$ so that $f_j \to f$ in the L^p norm.

Our hypothesis tells us that, if $\epsilon > 0$, then there is a number $J > 0$ so that if $j, k > J$, then

$$\int |f_j - f_k|^p \, d\mu = \|f_j - f_k\|_{L^p}^p < \epsilon^p. \tag{5.13.1}$$

Thus there exists a subsequence $\{f_{j_k}\}$ such that

$$\|f_{j_{k+1}} - f_{j_k}\|_{L^p} < 2^{-k}$$

for $k \in \mathbb{N}$. Define

$$g(x) = |f_{j_1}(x)| + \sum_{k=1}^{\infty} |f_{j_{k+1}}(x) - f_{j_k}(x)|. \tag{5.13.2}$$

Thus g is measurable, nonnegative, and integrable.

By Fatou's lemma,

$$\int |g|^p \, d\mu \le \liminf_{n \to \infty} \int \left\{ |f_{j_1}| + \sum_{j=1}^{n} |f_{j_{k+1}} - f_{j_k}| \right\}^p d\mu.$$

Taking pth roots of both sides and applying Minkowski's inequality we find that

$$\left\{ \int |g|^p \, d\mu \right\}^{1/p} \le \liminf_{n \to \infty} \left\{ \|f_{j_1}\|_{L^p} + \sum_{k=1}^{n} \|f_{j_{k+1}} - f_{j_k}\|_{L^p} \right\}$$

$$\le \|f_{j_1}\|_{L^p} + 1.$$

Thus, if $E = \{x \in \mathbb{R} : g(x) < +\infty\}$, then $E \in \mathcal{X}$ and $\mu(X \setminus E) = 0$. Therefore the series in (5.13.2) converges almost everywhere and $g\chi_E$ belongs to L^p.

We now define f on $\mathbb{R}$ by

$$f(x) = \begin{cases} f_{j_1}(x) + \sum_{k=1}^{\infty} \left\{ f_{j_{k+1}}(x) - f_{j_k}(x) \right\} & \text{if} \quad x \in E, \\ 0 & \text{if} \quad x \notin E. \end{cases}$$

Since

$$|f_{j_k}| \le |f_{j_1}| + \sum_{\ell=1}^{k} |f_{j_{\ell+1}} - f_{j_\ell}| \le g,$$

and since $\{f_{j_k}\}$ converges almost everywhere to f, the dominated convergence theorem now tells us that $f \in L^p$. Also, since $|f - f_{j_k}|^p \le 2^p \cdot g^p$, we conclude from the dominated convergence theorem that $0 = \lim_{k \to \infty} \|f - f_{j_k}\|_{L^p}$ so that $\{f_{j_k}\}$ converges in the L^p norm to f.

Because of (5.13.1), if $\epsilon > 0$ and $m > J$ as at the start of the proof, and if k is sufficiently large, then

$$\int |f_m - f_{j_k}|^p \, d\mu < \epsilon^p.$$

Applying Fatou's lemma, we now conclude that

$$\int |f_m - f|^p \, d\mu \le \liminf_{k \to \infty} \int |f_m - f_{j_k}|^p \, d\mu \le \epsilon^p$$

whenever $m > J$. This proves that the sequence $\{f_j\}$ converges to f in the L^p norm. $\qquad\square$

Thus we now know that each L^p space, $1 \le p < \infty$, is a Banach space.

5.2   The Case p = ∞

When discussing L^p spaces above, we always restricted p to be less than $+\infty$. But in fact L^∞ is an interesting and important space in its own right. We consider it now.

First a little review. Let $f : X \to \mathbb{R}$ be a function. A real number a is called an *upper bound* for f if $f(x) \leq a$ for all x. A convenient way of saying this is that

$$f^{-1}((a, \infty)) = \{x \in X : f(x) > a\} = \emptyset.$$

Set

$$U_f = \{a \in \mathbb{R} : f^{-1}((a, \infty)) = \emptyset\}$$

and define

$$\sup f = \inf U_f.$$

This defines the supremum or least upper bound of the function f.

Now, by analogy, consider a measure space $(X, \mathcal{X}, \mu)$. Let f be a measurable function on X. A real number a is called an *essential upper bound* for f if the set $f^{-1}((a, \infty))$ has measure zero. In other words, a is an essential upper bound if $f(x) \leq a$ for almost all $x \in X$. As we did in the last paragraph, let

$$U_f^{\mathrm{ess}} = \{a \in \mathbb{R} : \mu(f^{-1}((a, \infty))) = 0\}$$

be the set of all essential upper bounds. We define the *essential supremum* of f to be

$$\mathrm{ess}\sup f = \inf U_f^{\mathrm{ess}}$$

if $U_f^{\mathrm{ess}} \neq \emptyset$ and $\mathrm{ess}\sup f = +\infty$ otherwise.

Of course the *essential infimum* of f is defined in just the same way. A function $f : X \to \mathbb{R}$ is said to be *essentially bounded* if it has a finite essential supremum and a finite essential infimum.

EXAMPLE 5.14 We work with Lebesgue measure as usual. Let

$$f(x) = \begin{cases} 6 & \text{if} & x = 3, \\ -5 & \text{if} & x = -3, \\ 1 & \text{if} & x \neq 3, -3. \end{cases}$$

The supremum of this function is 6 and the infimum is -5. But these values are assumed only on a subset of the domain that has measure 0. Off that set of measure 0, the function is constantly equal to 1. Thus the essential infimum and the essential supremum of f are both 1.

Now define

$$g(x) = \begin{cases} x^5 & \text{if} & x \in \mathbb{Q}, \\ \arctan 2x & \text{if} & x \in \mathbb{R} \setminus \mathbb{Q}. \end{cases}$$

This function is unbounded, both above and below. So its supremum is $+\infty$ and its infimum is $-\infty$. But the unboundedness is based on values the function takes on domain elements in the rational numbers (which is a set of measure 0). On the irrational numbers, which is a set of full measure, the function is bounded above by $\pi/2$ and below by $-\pi/2$. Thus the essential supremum is $\pi/2$ and the essential infimum is $-\pi/2$.

Definition 5.15 Let $(X, \mathcal{X}, \mu)$ be a measure space. The space $L^\infty(X, \mu)$ or just L^∞ is defined to be the collection of essentially bounded functions. The norm on L^∞ is

$$\|f\|_{L^\infty} = \max\{|\text{ess sup } f|, |\text{ess inf } f|\}.$$

Theorem 5.16 *The space L^∞ is a Banach space.*

Proof: That L^∞ is a linear space is routine, and we leave the details for the reader.

The interesting part is to check that L^∞ is complete. So let $\{f_j\}$ be a Cauchy sequence in L^∞. Let $E \subseteq X$ be a set of measure 0 such that $|f_j(x)| \le \|f_j\|_{L^\infty}$ for $x \notin E$ and $j = 1, 2, \ldots$ and also so that $|f_j(x) - f_k(x)| \le \|f_j - f_k\|_{L^\infty}$ for all $x \notin E$, $j, k = 1, 2, \ldots$ (note that the set E is obtained by taking the countable union of sets of measure zero coming from different values of j and k). Then the sequence $\{f_j\}$ is uniformly convergent on $X \setminus E$. We let

$$f(x) = \begin{cases} \lim_{j \to \infty} f_j(x) & \text{if} \quad x \notin E, \\ 0 & \text{if} \quad x \in E. \end{cases}$$

It follows that f is measurable. It is easy to check that $\lim_{j \to \infty} \|f_j - f\|_{L^\infty} \to 0$. Thus L^∞ is complete. $\qquad\square$

Exercises

1. Let $C([0, 1])$ be the continuous functions on the interval $[0, 1]$. This is a linear space. Define a norm for $f \in C([0, 1])$ by

$$\|f\|_{\text{sup}} = \max_{x \in [0,1]} |f(x)|.$$

   Prove that this is a Banach space.

2. Let $C([0, 1])$ be as in Exercise 1. Define a new norm by

$$\|f\|_0 = |f(0)|.$$

   Prove that this is a seminorm.

3. Let $C([0,1])$ be as in Exercise 1. Define a new norm by

$$\|f\|_1 = \int_0^1 |f(x)|\, dx\,.$$

Show that this is *not* a Banach space.

4. Let X be a normed linear space with norm $\|\ \|$. Define a distance d on elements $x, y \in X$ by $d(x,y) = \|x - y\|$. Prove that d is a metric.

5. Let $(X, \mathcal{X}, \mu)$ be a measure space. Let f be an L^p function, $1 \le p < \infty$. Let $\epsilon > 0$. Show that there is a simple function s so that $\|f - s\|_{L^p} < \epsilon$. What happens when $p = \infty$?

6. Let $f : [1, \infty) \to \mathbb{R}$ be given by $f(x) = 1/x$. Show, using Lebesgue measure, that $f \notin L^1$, but $f \in L^p$ for $1 < p \le \infty$.

7. Fix $1 < p_0 < \infty$. Produce a function f on the interval $[0,1]$ such that $f \in L^p$ if and only if $1 \le p < p_0$.

8. Use Lebesgue measure on $[0,1]$. Let $j > 1$. Let f be a measurable function and set $E_j = \{x \in \mathbb{R} : j - 1 \le |f(x)| < j\}$. Show that $f \in L^1([0,1])$ if and only if

$$\sum_{j=1}^{\infty} j\mu(E_j) < +\infty\,.$$

If $p > 1$, show that $f \in L^p$ if and only if

$$\sum_{j=1}^{\infty} j^p \mu(E_j) < +\infty\,.$$

9. Use Lebesgue measure on $[0,1]$. Let $p > 1$. Prove that, if $f \in L^p([0,1])$, then $f \in L^r([0,1])$ for $1 \le r \le p$.

10. Is there an analogue to the result of Exercise 9 on the interval $[1, \infty)$?

11. Use Lebesgue measure. Let $f \in L^p(\mathbb{R})$, $1 \le p < +\infty$. Let $\epsilon > 0$. Show that there is a set $E \subseteq \mathbb{R}$ with $\mu(E) < \infty$ such that, if $F \subseteq \mathbb{R}$ with $F \cap E = \emptyset$, then $\|f \cdot \chi_F\|_{L^p} < \epsilon$.

12. Let $(X, \mathcal{X}, \mu)$ be a measure space. Show that the space L^∞ is contained in L^1 if and only if X has finite measure. Prove that, if $\mu(X) = 1$ and $f \in L^\infty$, then

$$\|f\|_{L^\infty} = \lim_{p \to \infty} \|f\|_{L^p}\,.$$

13. Let $p \ge 1$. For $\mathbf{x} = (x_1, x_2, \ldots, x_N) \in \mathbb{R}^n$, define

$$\|\mathbf{x}\| = (x_1^p + x_2^p + \cdots x_N^p)^{1/p}\,.$$

Show that this is a norm. In particular, prove the triangle inequality.

6

The Concept of Outer Measure

6.1 Outer Lebesgue Measure

For this and the next three chapters we shall concentrate on presenting the idea of Lebesgue measure from a different point of view. This is using the idea of outer measure. In fact outer measures are quite intuitive, and you may find this approach to be appealing. This is the approach that many texts use.

The approach that we present here works very well in $\mathbb{R}^N$ for all positive, integer values of N. But, in order to keep things simple, we shall concentrate on $\mathbb{R}^1$.

Definition 6.1 Let $E \subseteq \mathbb{R}$. We define the *Lebesgue outer measure $m^*(E)$* of E to be

$$m^*(E) = \inf \left\{ \sum_{j=1}^{\infty} \ell(I_j) \right\}, \tag{6.1.1}$$

where the infimum is taken over all sequences $\{I_j\}$ of open intervals in $\mathbb{R}$ that cover E in the sense that

$$E \subseteq \bigcup_{j=1}^{\infty} I_j. \tag{6.1.2}$$

Remark 6.2 Outer measure has these properties.

1. Since the intervals $I_j = (j-1, j+1)$ cover all of $\mathbb{R}$, they certainly cover any subset E of $\mathbb{R}$. So the infimum in (6.1.1) is *not* over the empty set. Clearly $m^*(E) \geq 0$. It is also certainly possible for $m^*(E) = +\infty$.

2. The terms $\ell(I_j)$ are all nonnegative. So the series

$$\sum_{j=1}^{\infty} \ell(I_j) \tag{6.2.1}$$

is either **(i)** absolutely convergent (in which case the value of the sum does not depend on the order of the intervals) or **(ii)** divergent, in which case the sum takes the value $+\infty$.

Remark 6.3   **(a)** It is most common to use open intervals in the definition of outer measure. But we could just as easily use closed intervals, and the resulting theory would be the same.

(b) It is also possible to use half-open intervals to define the outer measure. But there is no compelling reason to do so.

(c) Fix a number $\delta > 0$. We could define the outer measure using intervals that have length not exceeding δ. The same theory would result.

Theorem 6.4 *The outer measure function m^* satisfies:*

(a) $0 \leq m^*(E) \leq +\infty$ *for all* $E \subseteq \mathbb{R}$.

(b) $m^*(\emptyset) = 0$.

(c) *If* $E \subseteq F$, *then* $m^*(E) \leq m^*(F)$.

(d) *If* $\{E_j\}$ *are countably many subsets of* $\mathbb{R}$, *then*

$$m^*\left(\bigcup_{j=1}^{\infty} E_j\right) \leq \sum_{j=1}^{\infty} m^*(E_j).$$

Remark 6.5 Notice that we did *not* in this theorem specify countable additivity for outer measure. The simple reason is that outer measure is *not* countably additive when it is applied to *all* subsets of $\mathbb{R}$. We must find a criterion that allows us to restrict attention to a particular collection of subsets of $\mathbb{R}$. This is what we do below.

Proof of Theorem 6.4:

(a)   The first property is obvious from our earlier discussion.

(b)   This property is also obvious if we take each I_j to be the empty set.

(c)   If $\{I_j\}$ is a sequence of intervals whose union contain F then that union also contains E. That gives the result.

(d)   It suffices to prove the result when $m^*(E_k) < \infty$ for each k. Let $\epsilon > 0$ and, for each $j \in \mathbb{N}$, choose a sequence $\{I_j^k\}$ of intervals such that

$$E_k \subseteq \bigcup_{j=1}^{\infty} I_j^k \quad \text{and} \quad \sum_{j=1}^{\infty} \ell(I_j^k) \leq m^*(E_k) + \frac{\epsilon}{2^k}.$$

Since $\{I_j^k : j, k \in \mathbb{N}\}$ is a countable family of intervals that covers the set $\cup_{j=1}^{\infty} E_j$, we see from the definition of m^* that

$$
\begin{aligned}
m^*\left(\bigcup_{\ell=1}^{\infty} E_\ell\right) &\leq \sum_{j,k=1}^{\infty} \ell(I_j^k) \\
&= \sum_{k=1}^{\infty} \sum_{j=1}^{\infty} \ell(I_j^k) \\
&\leq \sum_{k=1}^{\infty} (m^*(E_k) + \epsilon/2^k) \\
&= \sum_{k=1}^{\infty} m^*(E_k) + \epsilon.
\end{aligned}
$$

Since $m^*(E) \geq 0$ for any set $E \subseteq \mathbb{R}$, the change from a double sum to an iterated sum is justified. Now, since $\epsilon > 0$ is arbitrary, the proof of property **(d)** is complete.

$\square$

Property **(d)** of Theorem 6.4 is commonly referred to as the *countable subadditivity* property. One consequence of **(d)** is that, if A and B are *disjoint* sets then

$$m^*(A \cup B) \leq m^*(A) + m^*(B).$$

From previous experience, one might expect to have equality in this last displayed equation. However such a property does not hold. We can show that, if there is a positive distance between A and B, then equality obtains. But without such an artificial hypothesis, we will find that we need to restrict attention to a special class of sets in order to get equality.

Proposition 6.6 *Let A and B be disjoint subsets of $\mathbb{R}$ with*

$$dist(A, B) \equiv \inf\{|a - b| : a \in A, b \in B\} > 0.$$

Then

$$m^*(A \cup B) = m^*(A) + m^*(B).$$

Proof: We saw in the previous displayed equation that the result is true with $\leq$ replacing $=$. Thus it suffices to prove the reverse inequality under the hypothesis that $m^*(A \cup B) < +\infty$ and $\delta = dist(A, B) > 0$.

Let $\epsilon > 0$. Let $\{I_j\}$ be a covering of $A \cup B$ such that

$$\sum_{j=1}^{\infty} \ell(I_j) \leq m^*(A \cup B) + \epsilon.$$

As previously noted, we may assume that the intervals I_j each have length less than δ. Thus none of the I_j can contain both points of A and points of B.

As a result, we can divide the intervals I_j into three classes:

(i)   The intervals J_j that contain points of A;

(ii) The intervals K_j that contain points of B;

(iii) The intervals H_j that contain neither points in A nor points in B.

Thus we have

$$m^*(A) \le \sum_j \ell(J_j) \quad \text{and} \quad m^*(B) \le \sum_j \ell(K_j).$$

From this it follows that

$$
\begin{aligned}
m^*(A) + m^*(B) &\le \sum_j \ell(J_j) + \sum_j \ell(K_j) + \sum_j \ell(H_j) \\
&\le \sum_j \ell(I_j) \\
&\le m^*(A \cup B) + \epsilon.
\end{aligned}
$$

We conclude that $m^*(A) + m^*(B) \le m^*(A \cup B) + \epsilon$. Since $\epsilon > 0$ is arbitrarily small, we conclude that $m^*(A) + m^*(B) \le m^*(A \cup B)$, as was to be proved. $\square$

Next we show that, at least for intervals, the outer measure m^* gives no surprises.

Proposition 6.7 *If I is any open interval, then $m^*(I) = \ell(I)$.*

Proof: Since the sequence $\{I, \emptyset, \emptyset, \ldots\}$ is a covering of I, it follows that $m^*(I) \le \ell(I) + \ell(\emptyset) + \ell(\emptyset) + \cdots = \ell(I) + 0 + 0 + \cdots = \ell(I)$. That establishes one direction.

For the opposite inequality, let $\epsilon > 0$ and let $\{I_j\}_{j=1}^m$ be a covering of I by open intervals so that

$$\sum_{j=1}^{\infty} \ell(I_j) \le m^*(I) + \epsilon.$$

Let J be a closed interval contained in I and so that $\ell(I) - \epsilon < \ell(J)$. The Heine-Borel theorem then tells us that there is an $m \in \mathbb{N}$ such that $J \subseteq \cup_{j=1}^m I_j$. See Figure 6.1.

Of course the intervals I_j will in general have some overlap. Let $p_1, p_2, \ldots, p_k$ be the endpoints of the I_j in the natural order in which they occur on the real line. Consider any closed interval which has endpoints some sequential pair p_j, p_{j+1} and which is a subset of one of the I_j. Call those closed

FIGURE 6.1
An open covering.

intervals $K_1, K_2, \ldots, K_p$. Likewise let $J_1, J_2, \ldots, J_q$ be those closed intervals into which J is divided by the p_j. Then we have

$$\ell(J) \;=\; \sum_{k=1}^{q} \ell(J_k)$$

$$\leq \; \sum_{\ell=1}^{p} \ell(K_\ell)$$

$$\leq \; \sum_{r=1}^{m} \ell(I_r)$$

$$\leq \; m^*(I) + \epsilon.$$

We see that $\ell(I) \leq \ell(J) + \epsilon \leq m^*(I) + 2\epsilon$. Since $\epsilon > 0$ is arbitrarily small, we conclude that $\ell(I) \leq m^*(I)$.

We conclude then that $\ell(I) = m^*(I)$, as was to be proved.   □

Remark 6.8 A similar result can be proved for closed intervals, or for half-open intervals. We leave the details for the interested reader.

It is an easily established fact, and we leave the details for the reader, that outer measure is translation invariant. That is to say, if $E \subseteq \mathbb{R}$ is a set and $a \in \mathbb{R}$ and $E_a \equiv \{e + a : e \in E\}$, then $m^*(E) = m^*(E_a)$.

Exercises

1. Calculate the outer measure of the Cantor ternary set.

2. Show that there is a set, constructed in a manner analogous to that for the Cantor ternary set, that has positive outer measure.

3. What can you say about the outer measure of the nonmeasurable set that we constructed in Chapter 1?

4. Suppose that E is a Lebesgue measurable set that has Lebesgue measure 0. What can you say about the outer measure of E?

5. Suppose that E is a Lebesgue measurable set that has outer measure 0. Then what can you say about the Lebesgue measure of E?

6. Show that the countable union of sets with outer measure 0 still has outer measure 0. Show that this result fails for uncountable unions.

7. Let $\rho > 0$. Give an example of a totally disconnected set that has outer measure ρ. Here a set is totally disconnected if it has no nontrivial connected subsets.

8. Let E be a set with outer measure $\rho > 0$. Show that E can be written as a (possibly countably infinite) union of subsets E_j so that each E_j has dyadic length. Here by "dyadic" we mean a negative power of 2.

9. Show that outer measure is translation invariant, as explained at the end of the section.

7

What Is a Measurable Set?

7.1   Identifying Measurable Sets

In our first go-around, we decided what a measurable set was by fiat. More precisely, we defined the concept of σ-algebra and then declared, "This is the σ-algebra of measurable sets." Now, in our new development, the point of view is different. Here we will understand measurable sets by more of a discovery method. And the measurable sets defined with our new technique will have all the desirable properties of a measure—including countable additivity.

We begin by recalling the definition of σ-algebra . We shall make considerable use of this concept in our current discussion. Compare with our consideration of σ-algebras in Section 1.3.

Definition 7.1 Let X be any set. Then a family $\mathcal{X}$ of subsets of X is said to be a *σ-algebra* in X if these conditions are satisfied:

(i) $\emptyset$ and X both belong to $\mathcal{X}$;

(ii) if $E \in \mathcal{X}$, then the complement $X \setminus E$ also belongs to $\mathcal{X}$;

(iii) if $\{E_j\}_{j=1}^{\infty}$ is a sequence of sets in $\mathcal{X}$, then the union $\cup_{j=1}^{\infty} E_j$ also belongs to $\mathcal{X}$.

Remark 7.2 We note the following points.

(a) If $\mathcal{X}$ is a σ-algebra of subsets of X, then the intersection of a sequence of sets in $\mathcal{X}$ also belongs to $\mathcal{X}$. This is an immediate consequence of de Morgan's laws.

(b) If X is any set, then $\{\emptyset, X\}$ is a trivial example of a σ-algebra .

(c) If X is any set and $E \subseteq X$, then $\mathcal{X} \equiv \{\emptyset, E, {}^cE, X\}$ is a σ-algebra .

(d) If X is any set, then the power set $\mathcal{P}(X)$ is a σ-algebra .

(e) If X is any set and if $\mathcal{X}_1$ and $\mathcal{X}_2$ are σ-algebras of subsets of X, then $\mathcal{X}_1 \cap \mathcal{X}_2$ is also a σ-algebra .

Definition 7.3 Let X be a set and let $\mathcal{X}$ be a σ-algebra of subsets of X. Then an $\widehat{\mathbb{R}}$-valued function μ with domain $\mathcal{X}$ is said to be a *measure* provided that

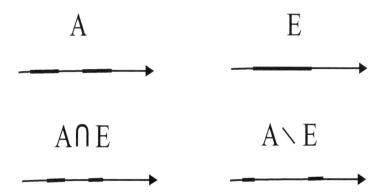

FIGURE 7.1
Carathéodory's condition.

   (i) $\mu(\emptyset) = 0$;

   (ii) $0 \le \mu(E) \le +\infty$ for all $E \in \mathcal{X}$;

   (iii) if $\{E_j\}_{j=1}^{\infty}$ is a sequence of sets in $\mathcal{X}$ that are pairwise disjoint, then

$$\mu\left(\bigcup_{j=1}^{\infty} E_j\right) = \sum_{j=1}^{\infty} \mu(E_j). \qquad (7.3.1)$$

EXAMPLE 7.4 If $X = \mathbb{N}$ and $\mathcal{X} = \{$all subsets of $X\}$, then define $\mu(E)$ to be the number of elements in E if E is a finite set and to be $+\infty$ if E is an infinite set. Then μ is a measure on $\mathcal{X}$. It is called the *counting measure* on $\mathbb{N}$.

Now we have reached a crucial juncture. We are going to explicitly define the criterion for measurability of a set.

Definition 7.5 Let m^* be the outer measure defined on all subsets of $\mathbb{R}$. A set $E \subseteq \mathbb{R}$ is said to satisfy the *Carathéodory condition* in case

$$m^*(A) = m^*(A \cap E) + m^*(A \setminus E) = m^*(A \cap E) + m^*(A \cap {}^c E) \qquad (7.5.1)$$

for all subsets $A \subseteq \mathbb{R}$. See Figure 7.1. The collection of all such sets will be denoted by $\mathcal{L}$.

We see that a set E satisfies Carathéodory's condition if E and its complement split every set A in an additive fashion. The sets that satisfy this condition are the ones that we shall think of as the *measurable sets*. The next result shows that the task of checking measurability can be simplified a bit.

Lemma 7.6 *A set E satisfies the Carathéodory condition if and only if, for each set A with $m^*(A) < \infty$, we have*

$$m^*(A) \geq m^*(A \cap E) + m^*(A \setminus E). \tag{7.6.1}$$

Proof: Since $A \cap E$ and $A \setminus E$ are disjoint and have union A, we see from Theorem 6.4(**d**) that we always have the inequality

$$m^*(A) \leq m^*(A \cap E) + m^*(A \setminus E).$$

Thus, if (7.6.1) is satisfied, then so is (7.5.1).

Observe in passing that (7.6.1) is trivial in case $m^*(A) = +\infty$. So it is only necessary to think about the case $m^*(A) < \infty$. □

Theorem 7.7 (Carathéodory) *Let m^* be the outer measure defined in Chapter 6. Then the set $\mathcal{L}$ of all subsets of $\mathbb{R}$ that satisfy the Carathéodory condition is a σ-algebra of subsets of $\mathbb{R}$. Furthermore, the restriction of m^* to $\mathcal{L}$ is a measure on $\mathcal{L}$.*

Proof: It is clear that the empty set $\emptyset$ satisfies (7.5.1). Also, if E satisfies (7.5.1), then so does its complement ^cE. Therefore the family of sets that satisfy the Carathéodory condition satisfies properties (**i**) and (**ii**) of Definition 7.1.

We next show that, if E and F satisfy (7.5.1), then so does $E \cap F$. This is the case because, since $E \in \mathcal{L}$, we have that

$$m^*(A) = m^*(A \cap E) + m^*(A \setminus E)$$

for any $A \subseteq \mathbb{R}$. Because $F \in \mathcal{L}$, we have

$$m^*(A \cap E) = m^*(A \cap E \cap F) + m^*(A \cap E \cap {}^cF).$$

Thus

$$m^*(A) = m^*(A \cap E \cap F) + m^*(A \cap E \cap {}^cF) + m^*(A \cap {}^cE).$$

But, since $E \in \mathcal{L}$, we have in addition that

$$\begin{aligned} m^*(A \cap {}^c(E \cap F)) &= m^*(A \cap {}^c(E \cap F) \cap E) + m^*(A \cap {}^c(E \cap F) \cap {}^cE) \\ &= m^*(A \cap {}^cF \cap E) + m^*(A \cap {}^cE). \end{aligned}$$

Thus we see that

$$m^*(A) = m^*(A \cap (E \cap F)) + m^*(A \cap {}^c(E \cap F))$$

for all sets A. Thus $E \cap F$ belongs to $\mathcal{L}$.

Since $\mathcal{L}$ contains the complements of sets in $\mathcal{L}$, we see from de Morgan's laws that, if $E, F \in \mathcal{L}$, then $E \cup F \in \mathcal{L}$. Furthermore, if $E \cap F = \emptyset$, then it follows from the fact that E satisfies (7.5.1) with A replaced by $A \cap (E \cup F)$ and $F = F \cap {}^c E$ that

$$
\begin{aligned}
m^*(A \cap (E \cup F)) &= m^*(A \cap (E \cup F) \cap E) + m^*(A \cap (E \cup F) \cap {}^c E) \\
&= m^*(A \cap E) + m^*(A \cap F).
\end{aligned}
$$

By induction we then see that, if $E_1, E_2, \ldots, E_k$ belongs to $\mathcal{L}$ and are pairwise disjoint, then $E_1 \cup E_2 \cup \cdots \cup E_k$ belongs to $\mathcal{L}$ and

$$
m^*(A \cap (E_1 \cup E_2 \cup \cdots \cup E_k)) = m^*(A \cap E_1) + \cdots + m^*(A \cap E_k)
$$

for all $A \subseteq \mathbb{R}$.

Now our task is to show that $\mathcal{L}$ is a σ-algebra and that m^* is countably additive on $\mathcal{L}$. To this end, let $\{E_j\}$ be a pairwise disjoint sequence of sets in $\mathcal{L}$ and let $E = \cup_{j=1}^{\infty} E_j$. Certainly $F_n \equiv \cup_{j=1}^{n} E_j$ belongs to $\mathcal{L}$ for all $n \in \mathbb{N}$. Further, if $A \subseteq \mathbb{R}$, then

$$
\begin{aligned}
m^*(A) &= m^*(A \cap F_n) + m^*(A \cap {}^c F_n) \\
&= m^* \left(\bigcup_{j=1}^{n} A \cap E_j \right) + m^*(A \cap {}^c F_n) \\
&= \sum_{j=1}^{n} m^*(A \cap E_j) + m^*(A \cap {}^c F_n).
\end{aligned}
$$

Since $F_n \subseteq E$, we see that $A \cap {}^c F_n \supseteq A \cap {}^c E$ for all $n \in \mathbb{N}$. Hence

$$
m^*(A) \geq \sum_{j=1}^{n} m^*(A \cap E_j) + m^*(A \cap {}^c E).
$$

This inequality implies that

$$
m^*(A) \geq \sum_{j=1}^{\infty} m^*(A \cap E_j) + m^*(A \cap {}^c E). \tag{7.7.1}
$$

It follows now from the countable subadditivity of m^* that

$$
m^*(A \cap E) = m^* \left(\bigcup_{j=1}^{\infty} A \cap E_j \right) \leq \sum_{j=1}^{\infty} m^*(A \cap E_j). \tag{7.7.2}
$$

Thus we have that

$$
m^*(A) \geq m^*(A \cap E) + m^*(A \cap {}^c E).
$$

This, in view of Lemma 7.6, implies that $E \in \mathcal{L}$. Hence $\mathcal{L}$ is a σ-algebra . What is more, if we take $A = E$ in (7.7.1) and (7.7.2), then we have

$$m^*(E) = \sum_{j=1}^{\infty} m^*(E_j) \,.$$

This shows that m^* is countably additive on $\mathcal{L}$. $\qquad\qquad\qquad\square$

Definition 7.8 If m^* is the outer measure defined in the last chapter, then the σ-algebra $\mathcal{L}$ of subsets of $\mathbb{R}$ that satisfy the Carathéodory condition is called the *Lebesgue σ-algebra* of $\mathbb{R}$. A set $E \in \mathcal{L}$ is called a *Lebesgue measurable subset of $\mathbb{R}$* or, briefly, a *measurable subset of $\mathbb{R}$*. The restriction of m^* to $\mathcal{L}$, which we now call m, is called the *Lebesgue measure* on $\mathbb{R}$.

Remark 7.9 Since m is the restriction of m^* to the σ-algebra $\mathcal{L}$, we know that $m(E) = m^*(E)$ for every $E \in \mathcal{L}$. Most of the time, when we know that a set E is measurable, we shall write $m(E)$ instead of $m^*(E)$.

Proposition 7.10 If I is an interval in $\mathbb{R}$, then I is measurable and $m(I) = \ell(I)$.

Proof: We shall give the proof for an open interval and leave the cases of the other intervals to the reader.

We saw in Lemma 7.6 that it is enough to show that, if $A \subseteq \mathbb{R}$ is such that $m^*(A) < +\infty$, then

$$m^*(A) \geq m^*(A \cap I) + m^*(A \setminus I) \,.$$

Let $n \in \mathbb{N}$ and let $I_n = \{x \in I : \mathrm{dist}(x, {}^cI) > 1/n\}$. Hence $I_n \subseteq I$. Also, since $I \setminus I_n$ lies in the union of 2 cells each of which has side length $1/n$, then $m^*(I \setminus I_n) \to 0$ as $n \to \infty$.

Notice that $A \supseteq (A \cap I_n) \cup (A \setminus I)$ and that $\mathrm{dist}(A \cap I_n, A \setminus I) \geq 1/n$. Thus we have from Proposition 6.6 that

$$\begin{aligned} m^*(A) &\geq m^*((A \cap I_n) \cup (A \setminus I)) \\ &= m^*(A \cap I_n) + m^*(A \setminus I) \,. \end{aligned} \qquad (7.10.1)$$

But we also know that

$$A \cap I = (A \cap I_n) \cup (A \cap (I \setminus I_n)) \,.$$

Thus it follows from the subadditivity and monotone character of m^* that

$$m^*(A \cap I_n) \leq m^*(A \cap I) \leq m^*(A \cap I_n) + m^*(I \setminus I_n) \,.$$

Thus we have

$$m^*(A \cap I) = \lim_{n \to \infty} m^*(A \cap I_n).$$

So, taking the limit in (7.10.1), we have

$$m^*(A) \geq m^*(A \cap I) + m^*(A \setminus I).$$

This shows, by Lemma 7.6, that I is a measurable set. $\qquad \square$

What we have accomplished thus far is both interesting and valuable. Namely, we have a measure m defined on a σ-algebra $\mathcal{L}$ of sets that agrees with the length function ℓ—and ℓ was originally defined only for intervals. Thus we have succeeded in extending ℓ to a notably larger collection of sets. We will spend some time seeing just how large $\mathcal{L}$ is.

It is conceivable that there is another measure defined on $\mathcal{L}$ that also agrees with ℓ on intervals. We now show that this is not the case.

Theorem 7.11 *If μ is a measure defined on the σ-algebra $\mathcal{L}$ that satisfies $\mu(I) = \ell(I)$ for all open intervals I, then $\mu = m$.*

Proof: For $n \in \mathbb{N}$, let $I_n = (-n, n)$. Let $E \in \mathcal{L}$ be any set with $E \subseteq I_n$ and let $\{J_k\}$ be a sequence of open intervals such that $E \subseteq \cup_{k=1}^\infty J_k$. Since μ is a measure and $\mu(J_k) = \ell(J_k)$ for all $k \in \mathbb{N}$, we see that

$$\mu(E) \leq \mu\left(\bigcup_{k=1}^\infty J_k \right) \leq \sum_{k=1}^\infty \mu(J_k) = \sum_{k=1}^\infty \ell(J_k).$$

Thus we have $\mu(E) \leq m^*(E) = m(E)$ for all measurable sets $E \subseteq I_n$.

Since μ and m are additive, we have

$$\mu(E) + \mu(I_n \setminus E) = \mu(I_n) = m(I_n) = m(E) + m(I_n \setminus E).$$

Because all these terms are finite and $\mu(E) \leq m(E)$ and $\mu(I_n \setminus E) \leq m(I_n \setminus E)$, we may conclude that $\mu(E) = m(E)$ for all measurable sets $E \subseteq I_n$.

Note that an arbitrary measurable set E can be written as the union of a disjoint sequence $\{E_j\}$ of sets, defined by

$$E_1 = E \cap I_1 \quad , \quad E_j = E \cap (I_j \setminus I_{j-1}), \quad \text{for } j > 1.$$

Since $\mu(E_j) = m(E_j)$ for all $j \in \mathbb{N}$, it follows that

$$\mu(E) = \sum_{j=1}^\infty \mu(E_j) = \sum_{j=1}^\infty m(E_j) = m(E).$$

In conclusion, μ and m agree on all measurable sets. $\qquad \square$

We wrap up this chapter with two useful and intuitively obvious facts about Lebesgue measure.

Theorem 7.12 *If E and F are Lebesgue measurable sets and if $E \subseteq F$, then $m(E) \leq m(F)$. If in addition $m(E) < +\infty$, then $m(F \setminus E) = m(F) - m(E)$.*

Proof: Since m is additive, and since we know that $F = E \cup (F \setminus E)$ and $E \cap (F \setminus E) = \emptyset$, then we have

$$m(F) = m(E) + m(F \setminus E).$$

Because $m(F \setminus E) \geq 0$, we conclude that $m(F) \geq m(E)$. If $m(E) < +\infty$, then we can subtract $m(E)$ from both sides of the above equation. $\square$

Theorem 7.13   (a) *If $\{E_j\}$ is an increasing sequence of Lebesgue measurable sets, then*

$$m \left(\bigcup_{j=1}^{\infty} E_j \right) = \lim_{n \to \infty} m(E_n). \tag{7.13.1}$$

(b) *If $\{F_j\}$ is a decreasing sequence of Lebesgue measurable sets and if $m(F_1) < \infty$, then*

$$m \left(\bigcap_{j=1}^{\infty} F_j \right) = \lim_{n \to \infty} m(F_n). \tag{7.13.2}$$

Proof:

(a) If $m(E_k) = +\infty$ for some $k \in \mathbb{N}$, then both sides of (7.13.1) are equal to $+\infty$. Thus we may assume that $m(E_j) < +\infty$ for all $j \in \mathbb{N}$. Let $A_1 = E_1$ and $a_j = E_j \setminus E_{j-1}$ for $j > 1$. Then $\{a_j\}$ is a pairwise disjoint sequence of measurable sets such that

$$E_j = \bigcup_{n=1}^{j} A_n \quad \text{and} \quad \bigcup_{j=1}^{\infty} E_j = \bigcup_{j=1}^{\infty} a_j.$$

Since m is countably additive, we see that

$$m \left(\bigcup_{j=1}^{\infty} E_j \right) = m \left(\bigcup_{n=1}^{\infty} A_n \right)$$

$$= \sum_{n=1}^{\infty} m(A_n)$$

$$= \lim_{p \to \infty} \left(\sum_{n=1}^{p} m(A_n) \right).$$

By Theorem 7.12, we see that

$$m(A_n) = m(E_n) - m(E_{n-1})$$

for $n > 1$. Hence the finite sum telescopes and

$$m\left(\bigcup_{j=1}^{\infty} E_j\right) = \lim_{n\to+\infty} m(E_n).$$

We conclude that (7.13.1) is proved.

(b) Let $E_j = F_1 \setminus F_j$ for $j \in \mathbb{N}$. Thus $\{E_j\}$ is an increasing sequence of measurable sets. If we apply part (a) of the present theorem, then we may infer that

$$
\begin{aligned}
m\left(\bigcup_{j=1}^{\infty} E_j\right) &= \lim_{j\to\infty} m(E_j) \\
&= \lim_{n\to\infty} [m(F_1) - m(F_n)] \\
&= m(F_1) - \lim_{n\to\infty} m(F_n).
\end{aligned}
$$

Since $\bigcup_{j=1}^{\infty} E_j = F_1 \setminus \bigcap_{n=1}^{\infty} F_n$, we may conclude from Theorem 7.12 that

$$m\left(\bigcup_{j=1}^{\infty} E_j\right) = m(F_1) - m\left(\bigcap_{n=1}^{\infty} F_n\right).$$

We conclude this discussion by combining the last two equations to obtain (7.13.2). □

Exercises

1. Prove that the interval $[0, 1]$ is Lebesgue measurable.

2. Prove that the Cantor ternary set is Lebesgue measurable.

3. Let $\mathcal{C}$ be the Cantor ternary set. Let $I_j = (2^{-j}, 2^{-j+1}]$ for $j = 1, 2, \ldots$. Set $C_j = I_j \cap \mathcal{C}$. Then certainly

$$\mathcal{C} = \{0\} \cup \bigcup_{j=1}^{\infty} C_j.$$

Verify countable additivity of Lebesgue measure in this instance.

4. Prove that there are uncountably many distinct sets of Lebesgue measure 0.

5. Prove that there are uncountably many distinct sets of Lebesgue measure 1.

6. Define a notion of inner measure by exhausting a given set E by compact sets. How does your idea of inner measure differ from the outer measure that we constructed in the text?

7. Prove the other cases of Proposition 7.10.

8. As enunciated in the text, prove that the intersection of two σ-algebras is a σ-algebra.

8

Decomposition Theorems

8.1 Signed Measures And the Hahn Decomposition

Definition 8.1 Let $\mathcal{X}$ be a σ-algebra on a set X. A function $\lambda : \mathcal{X} \to \mathbb{R}$ is called a *signed measure* if

(i) $\lambda(\emptyset) = 0$;

(ii) If E_j are measurable and pairwise disjoint then

$$\lambda \left(\bigcup_{j=1}^{\infty} E_j \right) = \sum_{j=1}^{\infty} \lambda(E_j).$$

We see that a signed measure is very much like a measure except that a measure is defined to take only positive values and 0 while a signed measure is allowed to also take negative values. Notices that we *do not* allow a signed measure to take values in the extended reals. We also remind the reader that condition **(ii)** is called countable additivity.

Definition 8.2 Let λ be a signed measure on the σ-algebra $\mathcal{X}$. A set $P \in \mathcal{X}$ is said to be *positive* with respect to λ if $\lambda(E \cap P) \geq 0$ for any $E \in \mathcal{X}$. A set $N \in \mathcal{X}$ is said to be *negative* with respect to λ if $\lambda(E \cap N) \leq 0$ for any $E \in \mathcal{X}$. A set $M \in \mathcal{X}$ is said to be a *null set* for λ if $\lambda(E \cap M) = 0$ for any set $E \in \mathcal{X}$.

Now we can prove an important decomposition theorem for signed measures.

Theorem 8.3 (Hahn Decomposition Theorem) *If λ is a signed measure on the σ-algebra $\mathcal{X}$ on the set X, then there exist sets P and N in $\mathcal{X}$ with $X = P \cup N$, $P \cap N = \emptyset$, and such that P is positive and N is negative with respect to λ.*

Proof: The class $\mathcal{P}$ of all positive sets is not empty, because it must at least contain the empty set $\emptyset$. Let

$$\alpha = \sup\{\lambda(A) : A \in \mathcal{P}\}.$$

Let $\{A_j\}$ be a sequence in $\mathcal{P}$ such that $\lim_{j\to\infty} \lambda(A_j) = \alpha$, and write $P = \cup_{j=1}^{\infty} A_j$.

Since the union of two positive sets is positive, the sequence $\{A_j\}$ can be chosen to be monotone increasing. We take this to be so. Clearly the P defined above is a positive set for λ because

$$\lambda(E \cap P) = \lambda\left(E \cap \bigcup_{j=1}^{\infty} A_j\right) = \lambda\left(\bigcup_{j=1}^{\infty}(E \cap A_j)\right) = \lim_{j\to\infty} \lambda(E \cap A_j) \geq 0\,.$$

Furthermore, $\alpha = \lim_{j\to\infty} \lambda(A_j) = \lambda(P) < \infty$.

We next prove that the set $N \equiv X \setminus P$ is a negative set. If not, then there is a measurable subset E of N so that $\lambda(E) > 0$. The set E cannot be a positive set, for if it were then $P \cup E$ would be positive with $\lambda(P \cup E) > \alpha$, contradicting the maximality of α. Hence E must itself contain sets with negative signed measure. Let n_1 be the least positive integer so that E contains a set E_1 in $\mathcal{X}$ such that $\lambda(E_1) \leq -1/n_1$. Now

$$\lambda(E \setminus E_1) = \lambda(E) - \lambda(E_1) > \lambda(E) > 0\,.$$

However $E \setminus E_1$ cannot be a positive set; if it were then $P_1 = P \cup (E \setminus E_1)$ would be a positive set with $\lambda(P_1) > \alpha$. Therefore $E \setminus E_1$ contains sets with negative signed measure.

Now let n_2 be the least positive integer so that $E \setminus E_1$ contains a set E_2 in $\mathcal{X}$ such that $\lambda(E_2) \leq -1/n_2$. As before, $E \setminus (E_1 \cup E_2)$ is not a positive set, and we next let n_3 be the least positive integer such that $E \setminus (E_1 \cup E_2)$ contains a set E_3 in $\mathcal{X}$ such that $\lambda(E_3) \leq -1/n_3$.

Repeating this argument, we obtain a disjoint sequence $\{E_j\}$ of measurable sets such that $\lambda(E_j) \leq -1/n_j$. Set $F = \cup_{j=1}^{\infty} E_j$ so that

$$\lambda(F) = \sum_{j=1}^{\infty} \lambda(E_j) \leq -\sum_{j=1}^{\infty} \frac{1}{n_j} \leq 0\,.$$

This shows that $1/n_j \to 0$.

If now G is a measurable subset of $E \setminus F$ and $\lambda(G) < 0$, then $\lambda(G) < -1/(n_j - 1)$ for sufficiently large j, contradicting the fact that n_j is the least positive integer such that $E \setminus (E_1 \cup \cdots \cup E_j)$ contains a set with signed measure less than $-1/n_j$. As a result, every measurable subset G of $E \setminus F$ must have $\lambda(G) \geq 0$. Hence $E \setminus F$ is a positive set for λ. Since

$$\lambda(E \setminus F) = \lambda(E) - \lambda(F) > 0\,,$$

we conclude that $P \cup (E \setminus F)$ is a positive set with signed measure exceeding α. Again, that is a contradiction.

It follows that the set $N = X \setminus P$ is a negative set for λ. Thus we have obtained the desired decomposition of X. $\square$

A pair of measurable sets P, N satisfying the conclusion of the Hahn decomposition theorem is said to be a *Hahn decomposition* of X with respect to λ. In general the Hahn decomposition is not unique. In applications this lack of uniqueness is not important.

Lemma 8.4 *If P_1, N_1 and P_2, N_2 are Hahn decompositions for X with respect to λ, and if $E \in \mathcal{X}$, then*

$$\lambda(E \cap P_1) = \lambda(E \cap P_2) \quad \text{and} \quad \lambda(E \cap N_1) = \lambda(E \cap N_2).$$

Proof: Since $E \cap (P_1 \setminus P_2)$ is contained in the positive set P_1 and also in the negative set N_2, we see that

$$\lambda(E \cap (P_1 \setminus P_2)) = 0.$$

Hence

$$\lambda(E \cap P_1) = \lambda(E \cap P_1 \cap P_2).$$

Similarly, one can show that

$$\lambda(E \cap P_2) = \lambda(E \cap P_1 \cap P_2).$$

In conclusion,

$$\lambda(E \cap P_1) = \lambda(E \cap P_2).$$

The result for N_1 and N_2 follows immediately. $\square$

Definition 8.5 Let λ be a signed measure on $\mathcal{X}$ and let P, N be a Hahn decomposition for λ. The *positive variation* for λ is the finite measure

$$\lambda^+(E) = \lambda(E \cap P).$$

Likewise the *negative variation* for λ is the finite measure

$$\lambda^-(E) = -\lambda(E \cap N).$$

The *total variation* of λ is the measure $|\lambda|$ which is defined for $E \in \mathcal{X}$ by

$$|\lambda|(E) = \lambda^+(E) + \lambda^-(E).$$

Note that we already encountered the ideas of positive and negative variation in Section 4.1.

Remark 8.6 It is a consequence of Lemma 8.4 that the positive and negative variations of λ are well defined and in fact do not depend on the Hahn decomposition.

It is also worth noting that

$$\lambda(E) = \lambda(E \cap P) + \lambda(E \cap N) = \lambda^+(E) - \lambda^-(E).$$

We now formalize the comments in this last remark.

Theorem 8.7 (Jordan Decomposition Theorem) *If λ is a signed measure on $\mathcal{X}$, then it is the difference of two finite measures on $\mathcal{X}$. That is to say, λ is the difference of λ^+ and λ^-. Furthermore, if $\lambda = \mu - \nu$ with μ, ν finite measures on $\mathcal{X}$, then*

$$\mu(E) \geq \lambda^+(E) \quad \text{and} \quad \nu(E) \geq \lambda^-(E)$$

for all $E \in \mathcal{X}$.

Proof: We have already proved that $\lambda = \lambda^+ - \lambda^-$. Since μ and ν have nonnegative values, we have

$$
\begin{aligned}
\lambda^+(E) &= \lambda(E \cap P) \\
&= \mu(E \cap P) - \nu(E \cap P) \\
&\leq \mu(E \cap P) \\
&\leq \mu(E).
\end{aligned}
$$

One shows similarly that $\lambda^-(E) \leq \nu(E)$. $\qquad\square$

Now we identify the positive and negative variations of a signed measure λ in a fashion similar to what we saw in Lemma 4.4.

Theorem 8.8 *If f is integrable with respect to the measure space $(X, \mathcal{X}, \mu)$, and if λ is defined by*

$$\lambda(E) = \int_E f \, d\mu,$$

then λ^+, λ^-, and $|\lambda|$ are given for $E \in \mathcal{X}$ by

$$\lambda^+(E) = \int_E f^+ \, d\mu \quad, \quad \lambda^-(E) = \int_E f^- \, d\mu,$$

$$|\lambda|(E) = \int_E |f| \, d\mu.$$

Proof: Let $Pf = \{x \in X : f(x) \geq 0\}$ and $N_f = \{x \in X : f(x) \leq 0\}$. Then $X = P_f \cup N_f$ and $P_f \cap N_f = \emptyset$. If $E \in \mathcal{X}$, then clearly $\lambda(E \cap P_f) \geq 0$ and $\lambda(E \cap N_f) \leq 0$. Hence P_f, N_f is a Hahn decomposition for λ. The result now follows. $\qquad\square$

8.2   The Radon–Nikodým Theorem

We begin with some terminology.

Definition 8.9 Let λ, μ be measures on a σ-algebra $\mathcal{X}$. We say that λ is *absolutely continuous* with respect to μ if, whenever $E \in \mathcal{X}$ and $\mu(E) = 0$, then $\lambda(E) = 0$. We then write $\lambda \ll \mu$.

Lemma 8.10 *Let $\mathcal{X}$ be a σ-algebra and let λ and μ be finite measures on $\mathcal{X}$. Then $\lambda \ll \mu$ if and only if, for each $\epsilon > 0$, there exists a $\delta > 0$ so that $E \in \mathcal{X}$ and $\mu(E) < \delta$ both imply that $\lambda(E) < \epsilon$.*

Proof: Suppose that $\lambda \ll \mu$. If, in addition, $\mu(E) = 0$, then $\lambda(E) < \epsilon$ for every $\epsilon > 0$. Therefore $\lambda(E) = 0$.

Conversely, suppose that there is an $\epsilon > 0$ and sets $E_j \in \mathcal{X}$ so that $\mu(E_j) < 2^{-j}$ and $\lambda(E_j) \geq \epsilon$. Let $F_j = \cup_{k=j}^{\infty} E_k$ so that $\mu(F_j) < 2^{-j+1}$ and $\lambda(F_j) \geq \epsilon$. Since $\{F_j\}$ is a decreasing sequence of measurable sets, we see that

$$\mu\left(\bigcap_{j=1}^{\infty} F_j\right) = \lim_{j \to \infty} \mu(F_j) = 0$$

and

$$\lambda\left(\bigcap_{j=1}^{\infty} F_j\right) = \lim_{j \to \infty} \lambda(F_j) \geq \epsilon.$$

We see then that λ is *not* absolutely continuous with respect to μ. $\square$

Theorem 8.11 (Radon–Nikodým) *Let λ and μ be σ-finite measures defined on a σ-algebra $\mathcal{X}$. Suppose that λ is absolutely continuous with respect to μ. Then there is a measurable function f on $\mathcal{X}$ such that*

$$\lambda(E) = \int_E f \, d\mu \quad \text{for all } E \in \mathcal{X}. \tag{8.11.1}$$

The function f is uniquely determined almost everywhere.

Proof: We shall first prove the result under the additional hypothesis that μ and λ are finite measures.

If $c > 0$, then let $P(c)$, $N(c)$ be a Hahn decomposition of X for the signed measure $\lambda - c\mu$. If $j \in \mathbb{N}$, then consider the measurable sets

$$A_1 = N(c), \quad A_{j+1} = N\big((j+1)c\big) \setminus \bigcup_{\ell=1}^{j} A_\ell.$$

Clearly the sets A_j, $j \in \mathbb{N}$, are disjoint and

$$\bigcup_{j=1}^{k} N(jc) = \bigcup_{j=1}^{k} A_j \,.$$

It follows that

$$A_j = N(jc) \setminus \bigcup_{\ell=1}^{j-1} N(\ell c) = N(jc) \cap \bigcap_{\ell=1}^{j-1} P(\ell c) \,.$$

As a result, if E is a measurable subset of A_j, then $E \subseteq N(jc)$ and $E \subseteq P((j-1)c)$ so that

$$(j-1)c\mu(E) \le \lambda(E) \le jc\mu(E) \,. \tag{8.11.2}$$

Define B by

$$B = X \setminus \bigcup_{j=1}^{\infty} A_j = \bigcap_{j=1}^{\infty} P(jc) \,.$$

Thus $B \subseteq P(jc)$ for all $j \in \mathbb{N}$. This tells us that

$$0 \le jc\mu(B) \le \lambda(B) \le \lambda(X) < +\infty$$

for all $j \in \mathbb{N}$, so that $\mu(B) = 0$. Since $\lambda \ll \mu$, we see that $\lambda(B) = 0$.

Now let f_c be defined by

$$f_c(x) = \begin{cases} (j-1)c & \text{if} \quad x \in A_j \\ 0 & \text{if} \quad x \in B \,. \end{cases}$$

If E is any measurable set, then E is the union of the disjoint sets $E \cap B$ and $E \cap A_j$ for $j \in \mathbb{N}$. Hence it follows from (8.11.2) that

$$\int_E f_c \, d\mu \le \lambda(E) \le \int_E (f_c + c) \, d\mu \le \int_E f_c \, d\mu + c\mu(X) \,.$$

We next employ the preceding construction with $c = 2^{-j}$, $j \in \mathbb{N}$, to obtain a sequence of functions denoted by f_j. Thus we have

$$\int_E f_j \, d\mu \le \lambda(E) \le \int_E f_j \, d\mu + 2^{-j}\mu(X) \,, \tag{8.11.3}$$

for all $j \in \mathbb{N}$. Let $k \ge j$ and note that

$$\int_E f_j \, d\mu \le \lambda(E) \le \int_E f_k \, d\mu + 2^{-k}\mu(X) \,,$$

$$\int_E f_k \, d\mu \le \lambda(E) \le \int_E f_j \, d\mu + 2^{-j}\mu(X) \,.$$

From this we see that

$$\left| \int_E (f_j - f_k)\, d\mu \right| \leq 2^{-j} \mu(X)$$

for all $E \in \mathcal{X}$.

If we let E range over the sets where the integrand is positive or negative and combine all these, we conclude that

$$\int |f_j - f_k|\, d\mu \leq 2^{-j+1} \mu(X)$$

whenever $k \geq j$. Thus $\{f_j\}$ converges both in mean and in measure to a function f. Since the f_j are nonnegative, we may deduce from Proposition 12.9 that f is nonnegative.

Furthermore,

$$\left| \int_E f_j\, d\mu - \int_E f\, d\mu \right| \leq \int_E |f_j - f|\, d\mu \leq \int |f_j - f|\, d\mu .$$

Hence we may conclude from (8.11.3) that

$$\lambda(E) = \lim_{j \to \infty} \int_E f_j\, d\mu = \int_E f\, d\mu$$

for all $E \in \mathcal{X}$. This completes the proof of the existence assertion of the theorem in the special case when λ and μ are finite measures.

We omit the proof of the more general case, but refer the reader to [1, pp. 86–87] for the details. Likewise for the uniqueness part of the theorem. $\square$

The function f whose existence is established in the theorem is usually called the *Radon-Nikodým derivative* of λ with respect to μ. It is denoted by $d\lambda/d\mu$. The function f is not necessarily integrable unless λ is finite.

We have seen that a measure λ is absolutely continuous with respect to a measure μ when sets which have small μ measure also have small λ measure. The dialectic opposite of "absolutely continuous" is "singular."

Definition 8.12 Two measures λ and μ on a σ-algebra $\mathcal{X}$ on a set X are said to be *mutually singular* if there are disjoint sets $A, B \in \mathcal{X}$ so that $X = A \cup B$ and $\lambda(A) = \mu(B) = 0$. In these circumstances we write $\lambda \perp \mu$.

Although this definition is plainly symmetric in λ and μ, it is still common to say that "λ is singular with respect to μ."

Theorem 8.13 (Lebesgue Decomposition Theorem) *Let λ and μ be σ-finite measures defined on a σ-algebra $\mathcal{X}$. There exists a measure λ_1 which is singular with respect to μ and another measure λ_2 which is absolutely continuous with respect to μ so that $\lambda = \lambda_1 + \lambda_2$. The measures λ_1 and λ_2 are unique.*

Proof: Let $\nu = \lambda + \mu$. Thus ν is a σ-finite measure. Since λ and μ are both absolutely continuous with respect to ν, the Radon-Nikodým theorem tells us that there are nonnegative, measurable functions f, g such that

$$\lambda(E) = \int_E f \, d\nu \quad , \quad \mu(E) = \int_E g \, d\nu$$

for all $E \in \mathcal{X}$. Let $A = \{x : g(x) = 0\}$ and $B = \{x : g(x) > 0\}$. We see immediately that $A \cap B = \emptyset$ and $X = A \cup B$. This A and B are the two sets that we seek.

Define, for $E \in \mathcal{X}$,

$$\lambda_1(E) = \lambda(E \cap A) \quad \text{and} \quad \lambda_2(E) = \lambda(E \cap B).$$

Because $\mu(A) = 0$, we see that $\lambda_1 \perp \mu$. To understand that $\lambda_2 \ll \mu$, notice that if $\mu(E) = 0$, then

$$\int_E g \, d\nu = 0.$$

Thus $g(x) = 0$ for ν-almost all $x \in E$. Hence $\nu(E \cap B) = 0$. Since $\lambda \ll \nu$,

$$\lambda_2(E) = \lambda(E \cap B) = 0.$$

Clearly $\lambda = \lambda_1 + \lambda_2$, so we have the required decomposition.

For the uniqueness, notice that if α is a measure such that $\alpha \ll \mu$ and if $\alpha \perp \mu$ also, then it must be that $\alpha = 0$. See also Exercise 7 at the end of the chapter.   □

8.3   The Riesz Representation Theorem

In this section we develop some ideas from functional analysis. In particular, we shall prove representation theorems for bounded linear functionals on L^p.

Definition 8.14 A *linear functional* on L^p is a mapping $\varphi : L^p \to \mathbb{R}$ which is linear.

The linear functional φ is *bounded* if there is a constant $M > 0$ such that

$$|\varphi(f)| \leq M \|f\|_{L^p}$$

for all $f \in L^p$. In this case, the *bound* or *norm* of the functional is defined to be

$$\|\varphi\| = \sup\{|\varphi(f)| : f \in L^p, \|f\|_{L^p} \leq 1\}.$$

EXAMPLE 8.15 Fix $1 \leq p < \infty$. Let $q = p/(p-1)$ and let $g \in L^q$. Define a linear functional φ on L^p by

$$\varphi(f) = \int f g \, d\mu \,.$$

Then φ is a linear functional with norm at most $\|g\|_{L^q}$. Just use Hölder's inequality to verify this assertion.

And in fact we leave it to the reader to check that the norm actually equals $\|g\|_{L^q}$. To see this, just assume $f \geq 0$ and let $f = g^{1/(p-1)}$. This will result in the equality $\varphi(f) = \|f\|_{L^p} \cdot \|g\|_{L^q}$.

The Riesz theorem gives a converse to the result in the example. We begin with a lemma. Note that a linear functional φ is called *positive* if $\varphi(f) \geq 0$ for all $f \in L^p$ such that $f \geq 0$.

Lemma 8.16 *Let φ be a bounded linear functional on L^p. Then there exist two positive bounded linear functionals φ^+ and φ^- such that*

$$\varphi(f) = \varphi^+(f) - \varphi^-(f)$$

for all $f \in L^p$.

Proof: If $f \geq 0$, then define

$$\varphi^+(f) = \sup\{\varphi(g) : g \in L^p, 0 \leq g \leq f\} \,.$$

Clearly $\varphi^+(cf) = c\varphi^+(f)$ for $c \geq 0$ and $f \geq 0$.

If $0 \leq g_j \leq f_j$ for $j = 1, 2$, then

$$\varphi(g_1) + \varphi(g_2) = \varphi(g_1 + g_2) \leq \varphi^+(f_1 + f_2) \,.$$

Taking suprema over all such $g_j \in L^p$, we find that

$$\varphi^+(f_1) + \varphi^+(f_2) \leq \varphi^+(f_1 + f_2) \,.$$

Conversely, if $0 \leq h \leq f_1 + f_2$, we let $g_1 = \max(h - f_2, 0)$ and $g_2 = \min(h, f_2)$. We infer then that $g_1 + g_2 = h$ and also $0 \leq g_j \leq f_j$ for $j = 1, 2$. Thus

$$\varphi(h) = \varphi(g_1) + \varphi(g_2) \leq \varphi^+(f_1) + \varphi^+(f_2)$$

for all $f_j \in L^p$ with $f_j \geq 0$.

If $f \in L^p$ is arbitrary, then we define

$$\varphi^+(f) = \varphi^+(f^+) + \varphi^+(f^-) \,,$$

where f^+ and f^- are respectively the positive and negative parts of f.

It is easy to see that φ^+ is a bounded linear functional on L^p. Next, for $f \in L^p$, we define

$$\varphi^-(f) = \varphi^+(f) - \varphi(f) \,.$$

We see that $\varphi \mapsto \varphi^-$ is plainly a bounded linear functional. From the definition of φ^+ we see that $\varphi \mapsto \varphi^-$ is also a positive linear functional. Finally, it
is obvious that $\varphi = \varphi^+ - \varphi^-$. $\qquad\qquad\qquad\qquad\qquad\qquad\qquad\qquad$ □

Theorem 8.17 (Riesz Representation Theorem, first version) *If* $(X,$
$\mathcal{X}, \mu)$ *is a σ-finite measure space and φ is a bounded linear functional on L^1,*
then there exists a function $g \in L^\infty$ such that the equation

$$\varphi(f) = \int fg \, d\mu$$

holds for all $f \in L^1$. Furthermore, $\|\varphi\| = \|g\|_{L^\infty}$. Also $g \geq 0$ if φ is a positive
linear functional.

Proof: We first treat the case that μ is a finite measure and φ is positive.
Define $\lambda : \mathcal{X} \to \mathbb{R}$ by

$$\lambda(E) = \varphi(\chi_E).$$

Clearly $\lambda(\emptyset) = 0$. If $\{E_j\}$ is an increasing sequence in $\mathcal{X}$ and $E = \cup_j E_j$,
then $\{\chi_{E_j}\}$ converges pointwise to χ_E. Since $\mu(X) < \infty$, it follows from the
Lebesgue Monotone Convergence Theorem that the same sequence converges
in L^1 to χ_E.
  Since

$$
\begin{aligned}
0 \ &\leq \ \lambda(E) - \lambda(E_j) \\
&= \ \varphi(\chi_E) - \varphi(\chi_{E_j}) \\
&= \ \varphi(\chi_E - \chi_{E_j}) \\
&\leq \ \|\varphi\| \cdot \|\chi_E - \chi_{E_j}\|_{L^1},
\end{aligned}
$$

we see (since $\lambda(E_j) \to \lambda(E)$) that λ is a measure. Furthermore, if $F \in \mathcal{X}$ and
$\mu(F) = 0$, then $\lambda(F) = 0$, hence $\lambda \ll \mu$.
  Applying the Radon-Nikodým theorem, we obtain a nonnegative, measurable function $g : X \to \mathbb{R}$ such that

$$\varphi(\chi_E) = \lambda(E) = \int \chi_E \cdot g \, d\mu$$

for all $E \in \mathcal{X}$. Linearity now implies that

$$\varphi(h) = \int h \cdot g \, d\mu$$

for all $\mathcal{X}$-measurable, simple functions h.
  If f is a nonnegative function in L^1, we let $\{h_j\}$ be a monotone increasing sequence of simple functions converging almost everywhere and in L^1 to

f. From the boundedness of φ, it is easily seen that $\varphi(f) = \lim_{j\to\infty} \varphi(h_j)$. Furthermore, the monotone convergence theorem tells us that

$$\varphi(f) = \lim_{j\to\infty} \int h_j g \, d\mu.$$

By linearity, this equality holds for arbitrary $f \in L^1$. That completes the case that μ is a finite measure.

We leave the case of σ-finite μ for the reader to consult in [1]. It is an exercise to check that $\|\varphi\| = \|g\|_{L^\infty}$. $\square$

What we have shown with this first version of the Riesz representation theorem is that the dual of the Banach space L^1 is L^∞. We conclude this chapter by now proving a version of the theorem for L^p.

Theorem 8.18 (Riesz Representation Theorem, second version) *Let $(X, \mathcal{X}, \mu)$ be an arbitrary measure space. Let φ be a bounded linear functional on L^p, $1 < p < \infty$. Then there exists a $g \in L^q$, $q = p/(p-1)$, so that*

$$\varphi(f) = \int fg \, d\mu$$

holds for all $f \in L^p$. Moreover, $\|\varphi\| = \|g\|_{L^q}$.

Proof: In the case that μ is finite, the proof of the preceding version of the Riesz theorem requires only minor changes to show that there exists a $g \in L^q$ with $\|\varphi\| = \|g\|_{L^q}$ and such that

$$\varphi(f) = \int fg \, d\mu$$

for all $f \in L^p$.

Now we complete the proof by noting that a bounded linear functional "vanishes off of a σ-finite set." Indeed, let $\{f_j\}$ be a sequence in L^p so that $\|f_j\| = 1$ and

$$\varphi(f_j) \geq \|\varphi\| \cdot \left(1 - \frac{1}{j}\right).$$

There exists a σ-finite set $X_0 \in \mathcal{X}$ outside of which all the f_j vanish. Let $E \in \mathcal{X}$ with $E \cap X_0 = \emptyset$. Then

$$\|f_j \pm t\chi_E\|_{L^p} \leq (1 + t^p \mu(E))^{1/p}$$

for $t \geq 0$ when $\mu(E) < \infty$.

Furthermore, since

$$\varphi(f_j) + \varphi(\pm t\chi_E) \leq |\varphi(f_j \pm t\chi_E)|,$$

we see that

$$|\varphi(t\chi_E)| \le \|\varphi\| \cdot \left\{(1 + t^p\mu(E))^{1/p} - \left(1 - \frac{1}{j}\right)\right\}$$

for all $j \in \mathbb{N}$.

First let $j \to \infty$ and then divide by $t > 0$ to obtain

$$|\varphi(\chi_E)| \le \|\varphi\| \cdot \frac{(1 + t^p\mu(E))^{1/p} - 1}{t}.$$

If we apply L'Hôspital's rule as $t \to 0^+$, we may conclude that $\varphi(\chi_E) = 0$ for any $E \in \mathcal{X}$, $\mu(E) < \infty$, outside of the σ-finite set X_0. Thus, if f is any function in L^p such that $X_0 \cap \{x \in X : f(x) \ne 0\}$, then it follows that $\varphi(f) = 0$.

Now we can apply the preceding argument to find a function g on X_0 which represents φ and we extend g to all of X by requiring that it vanish on the complement of X_0. This gives the desired function. $\qquad\square$

Exercises

1. Suppose that P is a positive set with respect to a signed measure λ. Let $E \in \mathcal{X}$ be such that $E \subseteq P$. Prove that E is positive with respect to λ.

2. Show that, if P_1 and P_2 are positive sets with respect to a signed measure λ, then also $P_1 \cup P_2$ is positive with respect to λ.

3. Let μ_1, μ_2, μ_3 be measures on the set X with σ-algebra $\mathcal{X}$. Show that if $\mu_1 \ll \mu_2$ and $\mu_2 \ll \mu_3$, then $\mu_1 \ll \mu_3$.

4. Let $\{\mu_j\}$ be a sequence of measures on the set X with σ-algebra $\mathcal{X}$ and $\mu_j(X) \le 1$ for each j. For $E \in \mathcal{X}$ define

$$\lambda(E) = \sum_{j=1}^{\infty} 2^{-j}\mu_j(E).$$

Show that λ is a measure and that $\mu_j \ll \lambda$ for each j.

5. Show that, if λ and μ are σ-finite and if $\lambda \ll \mu$, then the function f in the Radon-Nikodým theorem can be taken to be finite-valued on X.

6. Let λ, μ be σ-finite measures on the set X with σ-algebra $\mathcal{X}$. Assume that $\lambda \ll \mu$ and set $f = d\lambda/d\mu$. If g is a nonnegative, measurable function, then show that

$$\int g \, d\lambda = \int gf \, d\mu.$$

7. Let λ, μ be measures with $\lambda \ll \mu$ and $\lambda \perp \mu$. Show that $\lambda = 0$.

8. Let λ be a signed measure and μ a measure. Assume that $|\lambda| \perp \mu$. Show that λ^+ and λ^- are singular with respect to μ.

9. Show that the set of all signed measures on the set X with σ-algebra $\mathcal{X}$ forms a Banach space using the vector space operations

$$(c\mu)(E) = c\mu(E) \quad \text{and} \quad (\lambda + \mu)(E) = \lambda(E) + \mu(E)$$

and the norm $\|\mu\| = |\mu|(X)$ (where $|\mu|$ is the total variation of μ).

10. If g satisfies

$$\varphi(f) = \int fg \, d\mu$$

and

$$|\varphi(f)| \leq C\|f\|_{L^p}$$

for all $f \in L^p$, then show that $g \in L^q$ and $\|\varphi\| = \|g\|_{L^q}$.

11. Show that the Hahn decomposition is not unique.

9

Creation of Measures

9.1 Outer Measure

In the preceding chapters we have presented some quite elementary methods for creating measures. Now we introduce some more sophisticated techniques. And we also literally *construct* Lebesgue measure from the operation of measuring the lengths of intervals.

In fact we will be taking a new approach to the idea of measure. We begin by defining an algebra, which is something a bit weaker than a σ-algebra. Namely, a σ-algebra respects countable unions while an algebra only respects finite unions.

The good thing about an algebra is that it is very easy to specify a measure on an algebra. Then we explain how one can construct a σ-algebra that contains the algebra; finally we extend the measure to that σ-algebra. This is in fact how we will explicitly construct Lebesgue measure this time around.

Definition 9.1 We define the *length* of a bounded interval of the form

$$(a, b) \quad \text{or} \quad [a, b) \quad \text{or} \quad (a, b] \quad \text{or} \quad [a, b] \tag{9.1.1}$$

to be $b - a$. The length of an interval of the form

$$(-\infty, b] \quad \text{or} \quad (-\infty, b) \quad \text{or} \quad (b, \infty) \quad \text{or} \quad [b, \infty) \tag{9.1.2}$$

is the extended real number $+\infty$. We denote the length of a set S by $\ell(S)$. See Figures 9.1 and 9.2.

Definition 9.2 If we have finitely many pairwise disjoint intervals of the form (9.1.1) or (9.1.2), then the length of their union is defined to be the sum of the lengths of the component intervals. This aggregate length could be a finite, nonnegative number or it could be $+\infty$.

Now we define the notion of an algebra.

Definition 9.3 Let X be a given set. A family $\mathcal{A}$ of subsets of X is called an *algebra* or a *field* if

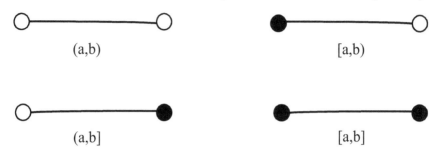

FIGURE 9.1
Lengths of intervals.

(i) $\emptyset$, X both belong to $\mathcal{A}$;

(ii) If E belongs to $\mathcal{A}$, then its complement $X \setminus E$ also belongs to $\mathcal{A}$;

(iii) If $E_1, E_2, \ldots, E_k$ belong to $\mathcal{A}$, then also their union $\cup_{j=1}^{k} E_j$ belongs to $\mathcal{A}$.

Now we do something a bit unusual. We define a measure on an algebra to be a scalar-valued function that respects countable unions.

Definition 9.4 Let $\mathcal{A}$ be an algebra of subsets of a set X. A *measure on $\mathcal{A}$* is a function $\mu : \mathcal{A} \to \mathbb{R}^+$ satisfying:

(a) $\mu(\emptyset) = 0$;

(b) $\mu(E) \geq 0$ for all $E \in \mathcal{A}$;

(c) If $\{E_j\}_{j=1}^{\infty}$ is a sequence of pairwise disjoint sets in $\mathcal{A}$ such that

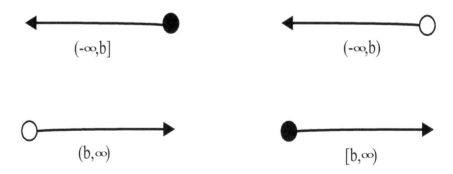

FIGURE 9.2
Intervals of infinite length.

FIGURE 9.3
The union of half-open intervals.

$\cup_{j=1}^{\infty} E_j$ also belongs to $\mathcal{A}$, then

$$\mu\left(\bigcup_{j=1}^{\infty} E_j\right) = \sum_{j=1}^{\infty} \mu(E_j).$$

Notice, in this definition, that we had to hypothesize that the union of the E_j lies in $\mathcal{A}$. That does *not* follow automatically from the definition of an algebra.

Lemma 9.5 *The collection $\mathcal{F}$ of all finite unions of sets of the forms (9.1.1) or (9.1.2) is an algebra of subsets of $\mathbb{R}$ and length is a measure on $\mathcal{F}$.*

Proof: It is apparent that $\mathcal{F}$ is an algebra. If we use ℓ to denote the length function, then 9.4(**a**) and 9.4(**b**) are obvious. To prove 9.4(**c**), it suffices to show that if one of the sets of the form (9.1.1) or (9.1.2) is the union of a countable collection of sets of this form, then the lengths add up correctly. We shall just treat an interval of the form $(a, b]$, leaving the other possibilities as an exercise for the reader.

Suppose now that

$$(a, b] = \bigcup_{j=1}^{\infty} (a_j, b_j], \tag{9.5.1}$$

where the intervals $(a_j, b_j]$ are pairwise disjoint. Refer to Figure 9.3. Let $(a_1, b_1], (a_2, b_2], \ldots, (a_k, b_k]$ be any finite collection of such intervals, and suppose that

$$a \le a_1 < b_1 \le a_2 < b_2 \le \cdots \le a_{k-1} < b_{k-1} \le a_k \le b_k \le b.$$

[We may need to renumber indices to achieve this goal, but that is simply a formality.] Now we have

$$\begin{aligned}
\sum_{j=1}^{k} \ell((a_j, b_j)) &= \sum_{j=1}^{k} (b_j - a_j) \\
&\le b_k - a_1 \\
&\le b - a \\
&= \ell((a, b]).
\end{aligned}$$

Since the index k is arbitrary, we conclude that

$$\sum_{j=1}^{\infty} \ell((a_j, b_j]) \leq \ell((a, b]). \tag{9.5.2}$$

Conversely, let $\epsilon > 0$ and let $\{\epsilon_j\}$ be a sequence of positive numbers with $\sum_j \epsilon_j < \epsilon/2$. Consider now the intervals

$$I_j = (a_j - \epsilon_j, b_j + \epsilon_n), \quad j \in \mathbb{N}.$$

From (9.5.1) we see that the open sets $\{I_j : j \in \mathbb{N}\}$ form a covering of the compact interval $[a, b]$. Hence there is a finite subcovering $I_1, I_2, \ldots, I_m$. By renumbering and discarding some possibly extra intervals, we may assume that

$$a_1 - \epsilon_1 < a \quad , \quad b < b_m + \epsilon_m ,$$

$$a_j - \epsilon_j < b_{j-1} + \epsilon_{j-1} , \quad j = 2, \ldots, m .$$

It follows from these inequalities that

$$\begin{aligned} b - a \quad &\leq \quad (b_m + \epsilon_m) - (a_1 - \epsilon_1) \\ &\leq \quad \sum_{j=1}^{m} [(b_j + \epsilon_j) - (a_j - \epsilon_j)] \\ &\leq \quad \sum_{j=1}^{m} (b_j - a_j) + \epsilon \\ &\leq \quad \sum_{j=1}^{\infty} (b_j - a_j) + \epsilon . \end{aligned}$$

Since $\epsilon > 0$ is arbitrary, we see that $\ell((a, b]) \leq \sum_{j=1}^{\infty} \ell((a_j, b_j])$. Combining this inequality with (9.5.2), we may conclude that the length function ℓ is countably additive on $\mathcal{F}$. $\qquad\square$

The next step in our program is to show that, if $\mathcal{A}$ is an algebra of subsets of a set X and if μ is a measure defined on $\mathcal{A}$, then there exists a σ-algebra $\widehat{\mathcal{A}}$ containing $\mathcal{A}$ and a measure $\widehat{\mu}$ defined on $\widehat{\mathcal{A}}$ so that $\widehat{\mu}(E) = \mu(E)$ for $E \in \mathcal{A}$. What we are saying then is that the measure μ can be extended from the algebra $\mathcal{A}$ to a measure on a σ-algebra $\widehat{\mathcal{A}}$ which contains $\mathcal{A}$.

Definition 9.6 Let $\mathcal{A}$ be an algebra of sets on X. Let μ be a measure on $\mathcal{A}$. If F is an arbitrary subset of X, then we define

$$\widehat{\mu}(F) = \inf \sum_{j=1}^{\infty} \mu(E_j), \tag{9.6.1}$$

where the infimum is extended over all sequences $\{E_j\}$ of sets in $\mathcal{A}$ such that

$$F \subseteq \bigcup_{j=1}^{\infty} E_j.$$ (9.6.2)

The function $\widehat{\mu}$ that we have just defined is called the *outer measure* generated by μ. This terminology is a bit confusing, just because $\widehat{\mu}$ is usually *not* a measure. It does, however, have some of the properties of a measure.

Lemma 9.7 *The function $\widehat{\mu}$ of Definition 9.6 has the following properties:*

(a) $\widehat{u}(\emptyset) = 0$;

(b) $\widehat{u}(F) \geq 0$ for $F \subseteq X$;

(c) If $F \subseteq G$, then $\widehat{u}(F) \leq \widehat{u}(G)$;

(d) If $F \in \mathcal{A}$, then $\widehat{u}(F) = \mu(F)$;

(e) If $\{F_j\}$ is a sequence of subsets of X, then

$$\widehat{u}\left(\bigcup_{j=1}^{\infty} F_j\right) \leq \sum_{j=1}^{\infty} \widehat{u}(F_j).$$

Proof: Statements (a), (b), and (c) are immediate consequences of Definition 9.6.

For (d), note that since $\{F, \emptyset, \emptyset, \ldots, \emptyset\}$ is a countable collection of sets in $\mathcal{A}$ whose union contain F, we see that

$$\widehat{\mu}(F) \leq \mu(F) + 0 + 0 + \cdot = \mu(F).$$

Conversely, if $\{E_j\}$ is any sequence of elements of $\mathcal{A}$ with $F \subseteq \cup_j E_j$, then $F = \cup_j (F \cap E_j)$. Since μ is assumed to be a measure on $\mathcal{A}$, we see that

$$\mu(F) \leq \sum_{j=1}^{\infty} \mu(F \cap E_j) \leq \sum_{j=1}^{\infty} \mu(E_j),$$

from which we infer that $\mu(F) \leq \widehat{\mu}(F)$.

To prove (e), let $\epsilon > 0$ and for each j choose a sequence $\{E_{jk}\}$ of sets in $\mathcal{A}$ so that

$$F_j \subseteq \bigcup_{k=1}^{\infty} E_{jk} \quad \text{and} \quad \sum_{k=1}^{\infty} \mu(E_{jk}) \leq \widehat{\mu}(F_j) + \frac{\epsilon}{2^j}.$$

Since $\{E_{jk} : j, k \in \mathbb{N}\}$ is a countable collection of elements of $\mathcal{A}$ whose union contains $\cup F_j$, it follows from the definition of $\widehat{\mu}$ that

$$\widehat{\mu}\left(\bigcup_{j=1}^{\infty} F_j\right) \leq \sum_{j=1}^{\infty} \sum_{k=1}^{\infty} \mu(E_{jk}) \leq \sum_{j=1}^{\infty} \widehat{\mu}(F_j) + \epsilon.$$

Since ϵ is arbitrary, we have obtained the desired inequality. $\square$

Property **(e)** of Lemma 9.7 is usually described by the phrase "$\widehat{\mu}$ is countably subadditive."

Remark 9.8 An important mathematical point must be made now. The outer measure $\widehat{\mu}$ assigns a length or measure to *every* subset of X. But of course we know from our considerations in Chapter 1 that that is not possible for a measure. So what is going on here?

The answer is that $\widehat{\mu}$ is *not* countably additive. In fact it is not necessarily finitely additive. What we shall do in our work below is to restrict $\widehat{u}$ to a smaller σ-algebra on which the outer measure *is* countably additive. The key result here is due to Constantin Carathéodory.

Definition 9.9 A subset E of X is said to be $\widehat{\mu}$-*measurable* if

$$\widehat{\mu}(A) = \widehat{\mu}(A \cap E) + \widehat{\mu}(A \setminus E) \tag{9.9.1}$$

for all subsets $A \subseteq X$. The collection of all $\widehat{\mu}$-measurable sets is denoted by $\widehat{\mathcal{A}}$.

Condition (9.9.1) defines an additivity property on $\widehat{\mu}$. Roughly speaking, a set E is $\widehat{\mu}$-measurable in case it and its complement are separated enough so that they divide an arbitrary set A in an additive fashion.

Theorem 9.10 (Carathéodory Extension Theorem) *The collection $\widehat{\mathcal{A}}$ of all $\widehat{\mu}$-measurable sets is a σ-algebra containing $\mathcal{A}$. Moreover, if $\{E_j\}$ is a pairwise disjoint sequence in $\widehat{\mathcal{A}}$, then*

$$\widehat{\mu}\left(\bigcup_{j=1}^{\infty} E_j\right) = \sum_{j=1}^{\infty} \widehat{\mu}(E_j). \tag{9.10.1}$$

Proof: Plainly $\emptyset$ and X are $\widehat{\mu}$-measurable. Also, if $E \in \widehat{\mathcal{A}}$, then the complement $X \setminus E$ is also in $\widehat{\mathcal{A}}$.

For our first step of the proof, we shall show that $\widehat{\mathcal{A}}$ is closed under intersection. Suppose that E, F are $\widehat{\mu}$-measurable. Then, for any $A \subseteq X$,

$$\widehat{\mu}(A \cap F) = \widehat{\mu}(A \cap F \cap E) + \widehat{\mu}((A \cap F) \setminus E).$$

Since $F \in \widehat{\mathcal{A}}$, we have

$$\widehat{\mu}(A) = \widehat{\mu}(A \cap F) + \widehat{\mu}(A \setminus F).$$

Let $B = A \setminus (E \cap F)$. Then clearly $B \cap F = (A \cap F) \setminus E$ and $B \setminus F = A \setminus F$. Since $F \in \widehat{\mathcal{A}}$, we see that

$$\widehat{\mu}(A \setminus (E \cap F)) = \widehat{\mu}((A \cap F) \setminus E) + \widehat{\mu}(A \setminus F).$$

Combining the last three equalities gives

$$\widehat{\mu}(A) = \widehat{\mu}(A \cap (E \cap F)) + \widehat{\mu}(A \setminus (E \cap F)).$$

This proves that $E \cap F \in \widehat{\mathcal{A}}$. Since $\widehat{\mathcal{A}}$ is closed under intersection and complementation, we may conclude that $\widehat{\mathcal{A}}$ is an algebra.

Now assume that $E, F \in \widehat{\mathcal{A}}$ and that $E \cap F = \emptyset$. If we take A to be instead $A \cap (E \cup F)$ in (9.9.1), we obtain

$$\widehat{\mu}(A \cap (E \cup F)) = \widehat{\mu}(A \cap E) + \widehat{\mu}(A \cap F).$$

Letting now $A = X$, we see that $\widehat{\mu}$ is additive on $\widehat{\mathcal{A}}$.

Now we show that $\widehat{\mathcal{A}}$ is a σ-algebra and that $\widehat{\mu}$ is countably additive on $\widehat{\mathcal{A}}$. Let $\{E_j\}_{j=1}^{\infty}$ be a pairwise disjoint collection in $\widehat{\mathcal{A}}$. Set $E = \cup_j E_j$. From the previous paragraph we know that $F_k = \cup_{j=1}^{k} E_j$ belongs to $\widehat{\mathcal{A}}$. Also, if A is any subset of X, then

$$\widehat{\mu}(A) = \widehat{\mu}(A \cap F_k) + \widehat{\mu}(A \setminus F_k) = \left[\sum_{j=1}^{k} \widehat{\mu}(A \cap E_j)\right] + \widehat{\mu}(A \setminus F_k).$$

Since $F_k \subseteq E$, we may see that $A \setminus E \subseteq A \setminus F_k$. Letting $k \to \infty$ we may conclude from the above that

$$\sum_{j=1}^{\infty} \widehat{\mu}(A \cap E_j) + \widehat{\mu}(A \setminus E) \leq \widehat{\mu}(A).$$

It follows now from Lemma 9.7(**e**) that

$$\widehat{\mu}(A \cap E) \leq \sum_{j=1}^{\infty} \widehat{\mu}(A \cap E_j)$$

and

$$\widehat{\mu}(A) \leq \widehat{\mu}(A \cap E) + \widehat{\mu}(A \setminus E).$$

Combining the last three inequalities yields

$$\widehat{\mu}(A) = \widehat{\mu}(A \cap E) + \widehat{\mu}(A \setminus E) = \sum_{j=1}^{\infty} \widehat{\mu}(A \cap E_j) + \widehat{\mu}(A \setminus E).$$

This shows in particular that $E = \cup_{j=1}^{\infty} E_j$ is $\widehat{\mu}$-measurable. Taking $A = E$, we obtain equation (9.10.1).

The last thing that we must do is to show that $\mathcal{A} \subseteq \widehat{\mathcal{A}}$. It was proved in Lemma 9.7(**d**) that, if $E \in \mathcal{A}$, then $\widehat{\mu}(E) = \mu(E)$. It remains to show that E is $\widehat{\mu}$-measurable. So let A be an arbitrary subset of X. Using Lemma 9.7(**e**), we see that

$$\widehat{\mu}(A) \leq \widehat{\mu}(A \cap E) + \widehat{\mu}(A \setminus E).$$

To prove the opposite inequality, let $\epsilon > 0$ and let $\{F_j\}$ be a sequence in $\mathcal{A}$ such that $A \subseteq \cup_j F_j$ and

$$\sum_{j=1}^{\infty} \mu(F_j) \leq \widehat{\mu}(A) + \epsilon.$$

Since $A \cap E \subseteq \cup_j (F_j \cap E)$ and $A \setminus E \subseteq \cup_j (F_j \setminus E)$, we may infer from Lemma 9.7(e) that

$$\widehat{\mu}(A \cap E) \leq \sum_{j=1}^{\infty} \mu(F_j \cap E) \quad \text{and} \quad \widehat{\mu}(A \setminus E) \leq \sum_{j=1}^{\infty} \mu(F_j \setminus E).$$

Thus we may conclude that

$$\begin{aligned}
\widehat{\mu}(A \cap E) + \widehat{\mu}(A \setminus E) \;\; &\leq \;\; \sum_{j=1}^{\infty} \Big\{ \mu(F_j \cap E) + \mu(F_j \setminus E) \Big\} \\
&= \;\; \sum_{j=1}^{\infty} \mu(F_j) \\
&\leq \;\; \widehat{\mu}(A) + \epsilon.
\end{aligned}$$

Since ϵ was arbitrary, we have proved the desired inequality and the set E belongs to $\widehat{\mathcal{A}}$. $\qquad\square$

Remark 9.11 The Carathéodory extension theorem shows that a measure μ on an algebra $\mathcal{A}$ can always be extended to a measure $\widehat{\mu}$ on a σ-algebra $\widehat{\mathcal{A}}$ containing $\mathcal{A}$. The σ-algebra $\widehat{\mathcal{A}}$ obtained in this way is automatically complete in the following sense. If $E \in \widehat{\mathcal{A}}$ with $\widehat{\mu}(E) = 0$ and if $B \subseteq E$, then $B \in \widehat{\mathcal{A}}$ and $\widehat{\mu}(B) = 0$.

To prove the last assertion, let A be an arbitrary subset of X and use Lemma 9.7(c) to note that

$$\widehat{\mu}(A) = \widehat{\mu}(E) + \widehat{\mu}(A) \geq \widehat{\mu}(A \cap B) + \widehat{\mu}(A \setminus B).$$

Now, as before, the inequality

$$\widehat{\mu}(A) \leq \widehat{\mu}(A \cap B) + \widehat{\mu}(A \setminus B)$$

follows from Lemma 9.7(e). Therefore B is $\widehat{\mu}$-measurable and

$$0 \leq \widehat{\mu}(B) \leq \widehat{\mu}(E) \leq 0.$$

Thus $\widehat{\mu}(B) = 0$ as desired.

Next we show that, in case μ is a σ-finite measure, then it has a unique extension to a measure on $\widehat{\mathcal{A}}$.

Theorem 9.12 (Hahn Extension Theorem) *Assume that μ is a σ-finite measure on an algebra $\mathcal{A}$. Then there exists a unique extension $\widehat{\mu}$ of μ to a measure on $\widehat{\mathcal{A}}$.*

Proof: The fact that $\widehat{\mu}$ is a measure on $\widehat{\mathcal{A}}$ was proved in Theorem 9.10—even without the σ-finiteness hypothesis. To prove the uniqueness, let ν be a measure on $\widehat{\mathcal{A}}$ which agrees with μ on $\mathcal{A}$.

Consider first the case that μ and hence $\widehat{\mu}$ are finite measures. Let $E \in \widehat{\mathcal{A}}$ and let $\{E_j\}_{j=1}^{\infty}$ be a collection of elements of $\mathcal{A}$ such that $E \subseteq \cup_j E_j$. Since ν is a measure and agrees with μ on $\mathcal{A}$, we see that

$$\nu(E) \leq \nu\left(\bigcup_{j=1}^{\infty} E_j\right) \leq \sum_{j=1}^{\infty} \nu(E_j) = \sum_{j=1}^{\infty} \mu(E_j).$$

Thus $\nu(E) \leq \widehat{\mu}(E)$ for any $E \in \widehat{\mathcal{A}}$. Since $\widehat{\mu}$ and ν are additive, we find that $\widehat{\mu}(E) + \widehat{\mu}(X \setminus E) = \nu(E) + \nu(X \setminus E)$. Since the terms on the righthand side are finite and not greater than the corresponding terms on the lefthand side, we conclude that $\widehat{\mu}(E) = \nu(E)$ for all $E \in \widehat{\mathcal{A}}$. This establishes the uniqueness when μ is a finite measure.

We leave the case of σ-finite μ for the reader to study in [1]. $\qquad\square$

9.2 Construction of Lebesgue Measure

Up until this point in the book we have taken the existence of Lebesgue measure—as an extension of the notion of length of an interval—for granted. Now we have the machinery developed to actually construct Lebesgue measure. And we do so.

Refer now to Lemma 9.5. We learned there that the collection $\mathcal{F}$ of all sets of the form (9.5.1) or (9.5.2) is an algebra of subsets of $\mathbb{R}$. Also the length function ℓ gives a measure on this algebra. If we apply the Carathéory extension theorem to this ℓ and $\mathcal{F}$, then we generate a measure space $(X, \mathcal{F}^*, \ell^*)$. The σ-algebra $\mathcal{F}^*$ is called the collection of *Lebesgue measurable sets* and the measure ℓ^* on $\mathcal{F}^*$ is called *Lebesgue measure*.

Although it is sometimes useful to work with $(X, \mathcal{F}^*, \ell^*)$, it is often more convenient to deal with the smallest σ-algebra containing $\mathcal{F}$ rather than with $\mathcal{F}^*$. And in fact that smallest σ-algebra is the collection $\mathcal{B}$ of Borel sets. The restriction of Lebesgue measure to the σ-algebra $\mathcal{B}$ is still called Lebesgue measure. It is a fact that every Lebesgue measurable set is contained in a Borel set with the same measure. And every Lebesgue measurable function is equal almost everywhere to a Borel measurable function. [We will verify these assertions below.] So there is little loss of generality to specialize down to the σ-algebra of Borel sets.

9.3   Borel-Stieltjes Measure

In this brief section we describe a generalization of the construction in the last section. The new measure that we construct here is useful in probability theory and other applications.

Let $g : \mathbb{R} \to \mathbb{R}$ be a monotone increasing function. We assume that g is right continuous at each point. We also suppose that $\lim_{x \to -\infty} g(x)$ and $\lim_{x \to +\infty} g(x)$ exist. The values of the latter two limits could in fact be $-\infty$, $+\infty$ respectively.

Now we define

$$
\begin{aligned}
\mu_g((a,b]) &= g(b) - g(a)\,, \\
\mu_g((-\infty, b]) &= g(b) - \lim_{x \to -\infty} g(x)\,, \\
\mu_g((a, +\infty)) &= \lim_{x \to +\infty} g(x) - g(a)\,, \\
\mu_g((-\infty, +\infty)) &= \lim_{x \to +\infty} g(x) - \lim_{x \to -\infty} g(x)\,.
\end{aligned}
$$

We go on to define μ_g on the algebra $\mathcal{F}$ (where $\mathcal{F}$ is as in the last section) of finite pairwise disjoint unions of such sets to be the corresponding sums. It is easy to see that this μ_g gives a σ-finite measure on the algebra $\mathcal{F}$. As a result, this measure has a unique extension, which we also denote by μ_g, to the algebra of all Borel subsets of $\mathbb{R}$. This extension is often called the *Borel-Stieltjes measure* generated by g. According to Theorem 9.10, μ_g actually has an extension to a complete σ-algebra which contains the Borel sets. This last extension is called the *Lebesgue-Stieltjes measure* generated by g. We encountered these ideas earlier in Section 2.1.

9.4   Linear Functionals on $C(X)$

Let X be a compact Hausdorff space. It is useful to know the dual of $C(X)$, the space of continuous functions on X under the supremum norm. We treat this matter in the present section. In fact it will turn out that the Banach space dual of $C([0,1])$ is a space of measures.

The main result here is, like two of our earlier results, called the Riesz representation theorem.

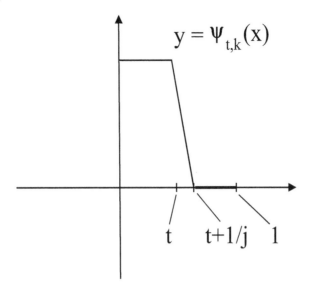

$$y = \Psi_{t,k}(x)$$

$t \qquad t+1/j \quad 1$

FIGURE 9.4
The function $\psi_{t,j}$.

Theorem 9.13 (Riesz Representation Theorem, third version) *Let $I =$
$[0, 1]$. Let φ be a positive, bounded linear functional on $C(I)$. Then there exists
a measure γ defined on the Borel subsets of I such that*

$$\varphi(f) = \int_I f \, d\gamma \tag{9.13.1}$$

for all $f \in C(I)$. Furthermore, the norm $\|\varphi\|$ of φ equals $\gamma(I)$.

Proof: If $t \in \mathbb{R}$ is such that $0 \le t < 1$ and j is a sufficiently large natural
number, then let $\psi_{t,j}$ be the function in $C(I)$ which equals 1 on $[0, t]$ and which
equals 0 on $(t + 1/j, 1]$. We also specify that $\psi_{t,j}$ be linear on $(t, t + 1/j)$. See
Figure 9.4.

If $j \le k$ and $x \in I$, then $0 \le \psi_{t,k}(x) \le \psi_{t,j}(x) \le 1$ so that the sequence
$\{\varphi(\psi_{t,j})\}$ is bounded and decreasing. If $t \in [0, 1)$, then we define

$$g(t) = \lim_{j \to \infty} \varphi(\psi_{t,j}).$$

Set $g(t) = 0$ for $t < 0$. Define $\psi_1(x) \equiv 1$ for all $x \in I$. If $t \ge 1$, we set
$g(t) = \varphi(\psi_1)$. Then g is a monotone increasing function on $\mathbb{R}$.

We claim that g is right continuous. This statement is clear if $t < 0$ or
$t \ge 1$. Let $t \in [0, 1)$ and $\epsilon > 0$. Choose

$$j > \max\{2, \|\varphi\| \cdot \epsilon^{-1}\}$$

to be so large that

$$g(t) \leq \varphi(\psi_{t,j}) \leq g(t) + \epsilon.$$

If ρ_j is the function in $C(I)$ which equals 1 on $[0, t + j^{-2}]$, which equals 0 on $(t + j^{-1} - j^{-2}, 1]$, and which is linear on

$$(t + j^{-2}, t + j^{-1} - j^{-2}],$$

then it is straightforward to check that $\|\rho_j - \psi_{t,j}\| \leq 1/j$. Therefore

$$\varphi(\rho_j) \leq \varphi(\psi_{t,j}) + \left(\frac{1}{j}\right)\|\varphi\| \leq g(t) + 2\epsilon.$$

It follows that $g(t) \leq g(t + j^{-2}) \leq g(t) + 2\epsilon$.

By the Hahn extension theorem, there exists a measure γ on the Borel subsets of $\mathbb{R}$ so that $\gamma([\alpha, \beta]) = g(\beta) - g(\alpha)$. This shows that $\gamma(E) = 0$ if $E \cap J = \emptyset$. It also implies that

$$\gamma([0, c]) = \gamma((-1, c]) = g(c)$$

and that $\|\varphi\| = |\varphi(\psi_1)| = g(1) = \gamma(I)$.

The last thing we must do is to show that equation (9.13.1) holds for $f \in C(I)$. Let $\epsilon > 0$. Since f is uniformly continuous on I, there is a $\delta > 0$ so that, if $|x - y| < \delta$ and $x, y \in I$, then $|f(x) - f(y)| < \epsilon$. Let

$$a = t_0 < t_1 < \cdots < t_m = b$$

be such that $\max\{t_k - t_{k-1}\} < \delta/2$. Choose j so large that $2/j < \min\{t_k - t_{k-1}\}$ and such that, for $k = 1, 2, \ldots, m$, we have

$$g(t_k) \leq \varphi(\psi_{t_k, j}) \leq g(t_k) + \epsilon(m\|f\|)^{-1}. \qquad (9.13.2)$$

Now we consider functions defined on I by

$$f_1(x) = f(t_1) \cdot \psi_{t_1, j}(x) + \sum_{k=2}^{\infty} f(t_k)\{\psi_{t_k, j}(x) - \psi_{t_{k-1}, j}(x)\},$$

$$f_2(x) = f(t_1)\chi_{[t_0, t_1]}(x) + \sum_{k=2}^{\infty} f(t_k)\chi_{(t_{k-1}, t_k]}(x).$$

Notice that $f_1 \in C(I)$ and that f_2 is a step function on I. Clearly $\sup_{x \in I}|f_2(x) - f(x)| \leq \epsilon$. As an exercise, the reader should also show that $\|f_1 - f\| \leq \epsilon$.

Thus we have

$$|\varphi(f) - \varphi(f_1)| \leq \epsilon\|\varphi\|.$$

In view of (9.13.2) we see that, if $2 \leq k \leq m$, then

$$|\varphi(\psi_{t_k, j} - \psi_{t_{k-1}, j}) - \{g(t_k) - g(t_{k-1})\}| \leq \epsilon(m\|f\|)^{-1}.$$

Now we apply φ to f_1 and integrate f_2 with respect to γ. The last inequality then gives us

$$\left|\varphi(f_1) - \int_I f_2 \, d\gamma\right| \le \epsilon.$$

Since f_2 lies within ϵ of f, we also have

$$\left|\int_I f_2 \, d\gamma - \int_I f \, d\gamma\right| \le \epsilon\gamma(I).$$

Combining the inequalities, we finally obtain

$$\left|\varphi(f) - \int_I f \, d\gamma\right| \le \epsilon(2\|\varphi\| + 1).$$

Because $\epsilon > 0$ is arbitrary, we finally deduce (9.13.1).   □

Remark 9.14 If the reader will check the proof of Theorem 9.13, it will be seen that an arbitrary bounded linear functional φ on $C(I)$ can be written as the difference $\varphi^+ - \varphi^-$ of two positive, bounded linear functionals. Using this observation, we can extend this latest Riesz representation theorem so that we may represent any bounded linear functional on $C(I)$ by means of integration with respect to a signed measure defined on the Borel subsets of I.

Exercises

1. Show that the family $\mathcal{G}$ of all finite unions of sets of the form

$$(a, b) \quad , \quad (-\infty, b) \quad , \quad (a, +\infty) \quad , \quad (-\infty, +\infty)$$

   is *not* an algebra (remember that an algebra is like a σ-algebra but it only respects finite unions—not countable unions). But the σ-algebra generated by $\mathcal{G}$ *is* the family $\mathcal{B}$ of Borel sets.

2. Prove that if the set $(a, +\infty)$ is the union of a pairwise disjoint sequence of sets $(a_j, b_j]$, then

$$\sum_{j=1}^{\infty} \ell((a_j, b_j]) = +\infty.$$

3. Let Y be the set of all rational numbers r with $0 < r \le 1$. Let $\mathcal{A}$ be the family of all finite unions of half-open intervals of the form

$$\{r \in Y : a < r \le b\}$$

or of the form
$$\{s \in Y : a \leq s < b\},$$

where $0 \leq a \leq b \leq 1$ and $a, b \in \mathbb{Q}$. Show that $\mathcal{A}$ is an algebra of subsets of Y. Prove also that every nonempty set in $\mathcal{A}$ is infinite.

4. Show that any countable subset of $\mathbb{R}$ has Lebesgue measure 0.

5. Give an example of an uncountable subset of $\mathbb{R}$ that has Lebesgue measure 0.

6. Give an example of a closed set in $\mathbb{R}$ which has positive Lebesgue measure, but whose interior has Lebesgue measure 0.

7. Let E be a Lebesgue measurable subset of $\mathbb{R}$ and let $\epsilon > 0$. Prove that there exists an open set $U \supseteq E$ so that
$$\ell^*(E) \leq \ell^*(U) \leq \ell^*(E) + \epsilon.$$

8. Let F be a Lebesgue measurable subset of $\mathbb{R}$ and let $\epsilon > 0$. Also assume that $F \subseteq I_j = (j, j+1]$ for some integer j. Prove that there exists a compact set $K \subseteq F$ so that
$$\ell^*(K) \leq \ell^*(F) \leq \ell^*(K) + \epsilon.$$

9. Let λ denote Lebesgue measure on $\mathbb{R}$. Let E be a Lebesgue measurable set with $\lambda(E) < +\infty$. If $\epsilon > 0$, then show that there exists an open set U which is the finite union of open intervals so that
$$\|\chi_E - \chi_U\| = \lambda((E \setminus U) \cup (U \setminus E)) \leq \epsilon.$$

Further show that, if $\epsilon > 0$, then there is a continuous function f such that
$$\|\chi_E - f\|_{L^1} = \int |\chi_E - f| \, d\lambda < \epsilon.$$

10. Use Lebesgue measure λ. Show that, if g is an L^1 function on the real line, then there is a continuous function f such that $\|g - f\|_{L^1} = \int |g - f| \, d\lambda < \epsilon$.

11. Let $\mathcal{B}$ be the σ-algebra of Borel sets in $\mathbb{R}$ and let λ denote Lebesgue measure on $\mathbb{R}$. Prove the following:

- $\lambda(U) > 0$ for every nonempty open set U;
- $\lambda(K) < +\infty$ for every compact set K;
- $\lambda(E + a) = \lambda(E)$, where $E \in \mathcal{B}$ and $E + a \equiv \{e + a : e \in E\}$.

12. Review the definition of μ_g in the text. Show that μ_g is a measure on the σ-algebra $\mathcal{F}$.

13. Complete the proof of Lemma 9.5 by treating the other cases.

10

Instances of Measurable Sets

10.1 Particular Sets

What we have done in the last several chapters is to construct a σ-algebra $\mathcal{L}$ of subsets of $\mathbb{R}$, called the *Lebesgue measurable sets.* And we have also constructed *Lebesgue measure,* which is a function whose domain is $\mathcal{L}$.

Certainly $\mathcal{L}$ includes the intervals, but in fact $\mathcal{L}$ is quite large and contains a great variety of sets. It is this last point that we address in the current chapter.

Lemma 10.1 *Let U be an open set in $\mathbb{R}$. Then U is the countable pairwise disjoint union of open intervals. See Figure 10.1*

Proof: Define a relation on U by $x \sim y$ if all the real numbers between x and y also lie in U. You can check that this is an equivalence relation.

Let E be an equivalence class from this relation. Then E is an interval. For if $a, b \in E$ then all the points between a and b must lie in E by the definition of the relation. And in fact E is an open interval by similar reasoning.

Of course the equivalence classes are pairwise disjoint. That ends the proof.
□

Proposition 10.2 *Every open set in $\mathbb{R}$ is Lebesgue measurable. Every closed set in $\mathbb{R}$ is Lebesgue measurable.*

Proof: We know that each open interval is Lebesgue measurable. And countable unions of measurable sets are measurable. Thus it follows from the preceding lemma that every open set is Lebesgue measurable.

That every closed set is Lebesgue measurable now follows by complementation.
□

FIGURE 10.1
The structure of an open set.

Definition 10.3 Let E be the intersection of a countable collection of open sets. Then E is called a G_δ.

Definition 10.4 Let F be the union of a countable collection of closed sets. Then F is called an F_σ.

EXAMPLE 10.5 Let $I_j = (-1/j, 1/j)$. Then each I_j is open, so $E = \cap_j I_j$ is a G_δ. But notice that $E = \{0\}$ is *not* open.

Let $J_j = [1/j, 1 - 1/j]$. Then each J_j is closed, so $F = \cup_j J_j$ is an F_σ. But notice that $F = (0, 1)$ is *not* closed.

EXAMPLE 10.6 A set that is the union of countably many G_δ sets is called a $G_{\delta\sigma}$ set. A set that is the intersection of countably many F_σ sets is called an $F_{\sigma\delta}$ set. Of course all these sets are Lebesgue measurable.

Definition 10.7 The smallest σ-algebra that contains all the open intervals is called the *Borel σ-algebra $\mathcal{B}$*.

We can think of the "smallest σ-algebra that contains all the open intervals" as the intersection of all the σ-algebras that contain the open intervals.

It is natural to observe that $\mathcal{B} \subseteq \mathcal{L}$. In fact the two σ-algebras are *not* equal, and we shall prove this assertion later in the chapter. The proof that we give will be abstract and non-constructive. But, at the very end of the book, we actually construct a Lebesgue measurable set that is not Borel.

10.2   Lebesgue Null Sets

We now treat sets that are small in the sense of Lebesgue measure but quite significant for our theory. These are the Lebesgue null sets.

Definition 10.8 A set $E \subseteq \mathbb{R}$ is called a *Lebesgue null set* if $m^*(E) = 0$.

EXAMPLE 10.9 Of course the empty set $\emptyset$ is a Lebesgue null set.

It is obvious that any finite set has measure zero.

In point of fact any countable set is a Lebesgue null set. To see this, let $\{a_j\}$ be a countable set. Let $\epsilon > 0$ and let

$$I_j = (a_j - 2^{-j-1}\epsilon, a_j + 2^{-j-1}\epsilon).$$

Then

$$E \subseteq \bigcup_{j=1}^{\infty} I_j$$

so that

$$m^*(E) \leq \sum_{j=1}^{\infty} m(I_j) \leq \sum_{j=1}^{\infty} 2^{-j}\epsilon = \epsilon.$$

Since this is true for any $\epsilon > 0$, we see that $m^*(E) = 0$.

It is common to summarize the last example by saying that any *denumerable set* has measure 0.

Proposition 10.10 *Any null set Z is Lebesgue measurable. It follows that $m(Z) = 0$. Furthermore, any subset of Z is Lebesgue measurable and a null set.*

Proof: Let $A \subseteq \mathbb{R}$ be any set. Since $Z \supseteq A \cap Z$ and $A \supseteq A \cap {}^c Z$, we may conclude from the monotonicity of m^* that

$$m^*(A) = m^*(Z) + m^*(A) \geq m^*(A \cap Z) + m^*(A \cap {}^c Z).$$

As a result, Lemma 7.6 tells us that Z is Lebesgue measurable. Therefore $m(Z) = m^*(Z) = 0$.

If W is a subset of Z, then $0 \leq m^*(W) \leq m^*(Z)$ so that W is also a null set. Hence W is Lebesgue measurable. □

The property that any subset of a Lebesgue null set is measurable and has measure zero is sometimes summarized by saying that Lebesgue measure is *complete*.

It is natural to think that a Lebesgue null set cannot have too many points. But this would be a misconception.

Proposition 10.11 *There exist Lebesgue null sets that have uncountably many points.*

Proof: Certainly the Cantor ternary set is a closed, indeed a compact, subset of $\mathbb{R}$ (see [4]). The construction of the Cantor set shows that the complement of the Cantor set in the unit interval has length 1. Hence the Cantor set is a Lebesgue null set. But we know that the Cantor set has uncountably many points. □

10.3   Invariance under Translation

An important feature of Lebesgue measure is that it is translation invariant. Intuitively speaking, this means that if we take a measurable set $E \subseteq \mathbb{R}$ and translate it by adding a constant a to each element of the set,

$$E_a = \{e + a : e \in E\},$$

then the measure of the set is unchanged. In other words, $m(E) = m(E_a)$. The present section develops this collection of ideas.

Proposition 10.12 *With E and E_a defined as above, $m(E) = m(E_a)$.*

Proof: Let A and B be arbitrary subsets of $\mathbb{R}$. Let $a \in \mathbb{R}$. Then it is easy to check that

$$A_a \cap B = [A \cap B_{-a}]_a .$$

Similarly, one can show that

$$[^c B]_a = {}^c [B_a] .$$

Now let $b = -a$. It follows from the obvious invariance of m^* under translation that

$$m^*(A_a \cap B) = m^*(A \cap B_b) . \tag{10.12.1}$$

We let $E \in \mathcal{L}$ and use equation (10.12.1) with $B = E$ and also with $B = {}^c E$ to obtain

$$
\begin{aligned}
m^*(A) &= m^*(A_a) \\
&= m^*((A_a) \cap E) + m^*((A_a) \cap {}^c E) \\
&= m^*(A \cap E_b) + m^*(A \cap ({}^c E)_b) \\
&= m^*(A \cap E_b) + m^*(A \cap {}^c (E_b))
\end{aligned}
$$

for all $A \subseteq \mathbb{R}$. Thus E_b is also Lebesgue measurable. Now it is straightforward to check that m^* is translation invariant—just because the notion of length of an interval is translation invariant. Thus we see that

$$m(E_b) = m^*(E_b) = m^*(E) = m(E) . \qquad \square$$

10.4   A Lebesgue Measurable Set That Is Not Borel

Our treatment of this topic will rely on Cantor's notion of cardinality, for which see [3].

Theorem 10.13 *There exists a Lebesgue measurable set of real numbers which is not Borel.*

Proof: Note that there are a countable number of open intervals in $\mathbb{R}$ with both endpoints rational. Call this cardinality $\mathcal{N}_0$. It is straightforward to see that $\mathcal{B}$ is the smallest σ-algebra that contains these intervals. Thus the cardinality of $\mathcal{B}$ is

$$(\mathcal{N}_0)^{\mathcal{N}_0} = c .$$

Here c is the cardinality of the real numbers $\mathbb{R}$.

Now we know that $\mathcal{L}$ contains a null set with uncountably many elements. Since an arbitrary subset of a null set is still a null set, we may conclude that $\mathcal{L}$ contains at least 2^c elements. Thus

$$\operatorname{card}(\mathcal{B}) = c < 2^c \leq \operatorname{card}(\mathcal{L}).$$

It follows then that $\mathcal{B}$ is a proper subset of $\mathcal{L}$.   □

Remark 10.14 It is actually possible to explicitly construct a Lebesgue measurable set that is not Borel. For the details, and some of the history, refer to [2].

In the very last section of this book we treat yet another method for constructing a Lebesgue measurable set that is not Borel.

Exercises

1. If f is a Lebesgue integrable function on the real line, then show that, for any fixed $a \in \mathbb{R}$,
    $$\int f(x)\, dx = \int f(x + a)\, dx.$$

2. Fix a Lebesgue integrable function φ on the real line. Define a linear operator
    $$T : f \longmapsto \int f(x - t)\varphi(t)\, dt$$
    for f integrable on the line. Show that this operator makes good sense. Further prove that T commutes with translations in a suitable sense.

3. Refer to Exercise 2 for terminology and notation. Show that
    $$Tf = \int f(t)\varphi(x - t)\, dt.$$
    Show that, if f is integrable, then Tf is integrable.

4. Refer to Exercise 2 for terminology and notation. Show that, if f^2 is integrable, then $(Tf)^2$ is integrable.

5. Show that there exists a function on the real line that is Lebesgue measurable but not Borel measurable.

6. Show that, if T is a translation invariant linear operator from the Lebesgue integrable functions to the Lebesgue integrable functions, then

$$Tf(x) = \int f(t)\varphi(x - t)\, dt$$

for some integrable function φ.

7. Refer to the construction of a nonmeasurable set in Section 1.2. Is that set Borel? Why or why not?

11

Approximation by Open And Closed Sets

The purpose of this chapter is to show that Lebesgue measurable sets are well approximated by open sets and by closed sets. Of course we have to specify the sense or the topology in which these approximations take place.

Lemma 11.1 Let $A \subseteq \mathbb{R}$. Let $\epsilon > 0$. Then there is an open set $U \subseteq \mathbb{R}$ such that $A \subseteq U$ and $m(U) \leq m^*(A) + \epsilon$. See Figure 11.1. Thus

$$m^*(A) = \inf\{m(U) : A \subseteq U, U \text{ open}\}. \tag{11.1.1}$$

Lemma 11.2 Let $A \subseteq \mathbb{R}$. Then there is a G_δ set E such that $A \subseteq E$ and $m^*(A) = m(E)$.

Proof of Lemma 11.1: We may assume that $m^*(A) < +\infty$. Of course we may require the intervals in the definition of outer measure to be open. So there exists a sequence $\{I_j\}$ of open intervals covering the set A such that

$$\sum_{j=1}^{\infty} \ell(I_j) \leq m^*(A) + \epsilon.$$

If we let $\mathcal{O} = \cup_{j=1}^{\infty} I_j$, then $\mathcal{O}$ is open and, by the countable subadditivity of m^* and Theorem 6.7, we see that

$$m(\mathcal{O}) \leq \sum_{j=1}^{\infty} m^*(I_j) = \sum_{j=1}^{\infty} \ell(I_j) \leq m^*(A) + \epsilon.$$

Equation (11.1.1) follows now from the definition of infimum. □

Proof of Lemma 11.2: For each $j \in \mathbb{N}$, let $\mathcal{O}_j$ be an open set such that $A \subseteq \mathcal{O}_j$ and $m(\mathcal{O}_j) \leq m^*(A) + 1/j$. Define $E = \cap_{j=1}^{\infty} \mathcal{O}_j$. Then E is a G_δ.

FIGURE 11.1
Approximation from the outside by an open set.

Thus $A \subseteq E \subseteq \mathcal{O}_j$ for every j. Also $m^*(A) \le m(E) \le m^*(A) + 1/j$ for all $j \in \mathbb{N}$. In conclusion, $m^*(A) = m(E)$. □

Corollary 11.3 *Any Lebesgue null set is a subset of a Borel null set.*

Proof: Let F be a Lebesgue null set. Then there is a G_δ set H such that $F \subseteq H$ and $m(H) = 0$. But then certainly H is a Borel set. □

It should be noted that, in Lemma 11.2, the set-theoretic difference $E \setminus A$ may not be a small set. We will see later that A is Lebesgue measurable if and only if $E \setminus A$ is a null set.

Proposition 11.4 *A set $E \subseteq \mathbb{R}$ is Lebesgue measurable if and only if, for every $\epsilon > 0$, there is an open set $\mathcal{O}$ such that $E \subseteq \mathcal{O}$ and $m^*(\mathcal{O} \setminus E) < \epsilon$.*

Proof: First we assume that E is measurable and $m(E) < \infty$. By Lemma 11.1, there is an open set $\mathcal{O}$ so that $E \subseteq \mathcal{O}$ and $m(\mathcal{O}) \le m(E) + \epsilon$. Since E is measurable and $E \subseteq \mathcal{O}$, we see that

$$m(\mathcal{O}) = m(\mathcal{O} \cap E) + m(\mathcal{O} \setminus E) = m(E) + m(\mathcal{O} \setminus E).$$

Because $m(E) < \infty$, we have

$$m(\mathcal{O} \setminus E) = m(\mathcal{O}) - m(E) < \epsilon.$$

In the case that $m(E) = +\infty$, we must work a bit harder. Let $E_1 = E \cap \{x : |x| \le 1\}$. If $j \ge 2$, let $E_j = E \cap \{x : j - 1 < |x| \le j\}$. For $j \in \mathbb{N}$, we let $\mathcal{O}_j$ be an open set with $E_j \subseteq \mathcal{O}_j$ and $m(\mathcal{O}_j \setminus E_j) < \epsilon/2^j$. Set $\mathcal{O} = \cup_{j=1}^\infty \mathcal{O}_j$. Then $\mathcal{O}$ is open, $E \subseteq \mathcal{O}$, and

$$\mathcal{O} \setminus E \subseteq \bigcup_{j=1}^\infty (\mathcal{O}_j \setminus E_j).$$

As a result, the countable subadditivity of m tells us that

$$m(\mathcal{O} \setminus E) \le \sum_{j=1}^\infty m(\mathcal{O}_j \setminus E_j) < \sum_{j=1}^\infty \frac{\epsilon}{2^j} = \epsilon.$$

Conversely, assume that for every $j \in \mathbb{N}$ there is an open set $\mathcal{O}_j \supseteq E$ so that $m^*(\mathcal{O}_j \setminus E) < 1/j$. Define $U = \cap_{j=1}^\infty \mathcal{O}_j$. Thus U is a G_δ set. So it is measurable. Also, since $U \subseteq \mathcal{O}_j$ for each j, we see that $U \setminus E \subseteq \mathcal{O}_j \setminus E$. Thus

$$0 \le m^*(U \setminus E) \le m^*(\mathcal{O}_j \setminus E) < \frac{1}{j}$$

for all $j \in \mathbb{N}$. Thus $m^*(U \setminus E) = 0$. And therefore $W = U \setminus E$ is a measurable set. And so $E = U \setminus W$ is a measurable set. □

Corollary 11.5 *Let $E \subseteq \mathbb{R}$ be a measurable set. For any $\epsilon > 0$ there then exists an open set $\mathcal{O} \supseteq E$ with $m(\mathcal{O}) \leq m(E) + \epsilon$. Thus*

$$m(E) = \inf\{m(\mathcal{O}) : \mathcal{O} \text{ is open and } \mathcal{O} \supseteq E\}.$$

Proof: Proposition 11.4 tells us that

$$m(\mathcal{O}) = m(E) + m(\mathcal{O} \setminus E) \leq m(E) + \epsilon. \qquad \square$$

We next consider the approximation of a Lebesgue measurable set from the outside by a G_δ. This turns out to be a useful characterization of Lebesgue measurability.

Proposition 11.6 *The following statements are logically equivalent for a set $E \subseteq \mathbb{R}$.*

(a) *The set $E \subseteq \mathbb{R}$ is Lebesgue measurable.*

(b) *There exists a G_δ set W with $E \subseteq W$ and $m^*(W \setminus E) = 0$.*

(c) *There exists a G_δ set W and a Lebesgue null set S so that $E \subseteq W$, $S \subseteq W$, and $E = W \setminus S$.*

Proof: This proof is a straightforward application of the preceding ideas. We leave the details to the reader. $\qquad \square$

11.1   Interior Approximation by Closed Sets

Now we consider the idea of approximating a Lebesgue measurable set from the inside by closed sets.

Theorem 11.7 *A set $S \subseteq \mathbb{R}$ is Lebesgue measurable if and only if, for each $\epsilon > 0$, there is a closed set $F \subseteq S$ with $m^*(S \setminus F) < \epsilon$.*

Proof: If S is measurable, then its complement ${}^c S$ is also measurable. By Proposition 11.4, there is an open set $\mathcal{O}$ with ${}^c S \subseteq \mathcal{O}$ and $m(\mathcal{O} \setminus {}^c S) < \epsilon$. Now we let $F = {}^c \mathcal{O}$. Thus F is closed. Also $F \subseteq S$. Further note that $S \setminus F = S \cap \mathcal{O} = \mathcal{O} \setminus {}^c S$. Hence

$$m(S \setminus F) = m(\mathcal{O} \setminus {}^c S) < \epsilon.$$

For the converse direction, note that for each $j \in \mathbb{N}$ there is a closed set $F_j \subseteq S$ with $m^*(S \setminus F_j) < 1/j$. Let $K = \cup_{j=1}^\infty F_j$. Then K is an F_σ set. So it is Lebesgue measurable. Since $F_j \subseteq K$, we see that $S \setminus K \subseteq S \setminus F_j$. Hence

$$m^*(S \setminus K) \leq m^*(S \setminus F_j) < \frac{1}{j}$$

FIGURE 11.2
Approximation from the inside by a compact set.

for all j. Thus $m^*(S \setminus K) = 0$. This implies that $W = S \setminus K$ is a measurable set. In conclusion, $S = K \cup W$ is measurable. □

Corollary 11.8 *Let* $S \subseteq \mathbb{R}$ *be Lebesgue measurable. Then, for any* $\epsilon > 0$, *there is a closed set* $F \subseteq S$ *with* $m(S) \leq m(F) + \epsilon$. *Thus we have*

$$m(S) = \sup\{m(F) : F \text{ closed and } F \subseteq S\}.$$

Proof: The set F in Theorem 11.7 is measurable and $F \subseteq S$. Thus

$$m(S) = m^*(S \cap F) + m^*(S \setminus F) \leq m(F) + \epsilon.$$ □

Next we characterize measurability in terms of approximation by F_σ sets from the inside.

Corollary 11.9 *The following statements are equivalent for a set* $E \subseteq \mathbb{R}$.

(a) *The set* E *is Lebesgue measurable.*

(b) *There exists an* F_σ *set* K *with* $K \subseteq E$ *and* $m^*(E \setminus K) = 0$.

(c) *There is an* F_σ *set* K *and a Lebesgue null set* W *such that* $K \subseteq E$, $W \subseteq E$, *and* $E = K \cup W$.

Proof: The proof is a straightforward application of the preceding ideas. We leave the details to the reader. □

11.2   Approximation from Inside by Compact Sets

Of course a compact set, being bounded, always has finite Lebesgue measure. So it is of interest that we can approximate *virtually any* Lebesgue measurable set from within by compact sets.

Theorem 11.10 *A set* $E \subseteq \mathbb{R}$ *with* $m^*(E) < \infty$ *is Lebesgue measurable if and only if, for each* $\epsilon > 0$, *there is a compact set* K *such that* $K \subseteq E$ *and* $m^*(E \setminus K) < \epsilon$. *See Figure 11.2.*

Proof: If E is measurable and $j \in \mathbb{N}$, then we let $E_j = E \cap \{x : |x| \leq j\}$. Since the sequence of sets E_j increases to E, we may conclude that the numerical sequence $\{m(E_j)\}$ increases to $m(E)$. Hence there is an index j_0 so that $m(E) < m(E_{j_0}) + \epsilon/2$. By Theorem 11.7, there is a closed set F with $F \subseteq E_{j_0}$ and $m(E_{j_0} \setminus F) < \epsilon/2$.

Of course E is the disjoint union of the sets $E \setminus E_{j_0}$ and E_{j_0} so we have

$$m(E) = m(E \setminus E_{j_0}) + m(E_{j_0}).$$

Since $m(E) < \infty$, we also know that

$$m(E \setminus E_{j_0}) = m(E) - m(E_{j_0}) < \frac{\epsilon}{2}.$$

Furthermore, $E \setminus F$ is the union of the disjoint sets $E \setminus E_{j_0}$ and $E_{j_0} \setminus F$. So we have

$$m(E \setminus F) = m(E \setminus E_{j_0}) + m(E_{j_0} \setminus F) < \epsilon.$$

Since $F \subseteq E_{j_0}$ is closed and bounded, it is compact. That proves the forward direction of the theorem.

For the converse, suppose that for each $j \in \mathbb{N}$ there is a compact set F_j with $F_j \subseteq E$ and $m^*(E \setminus F_j) < 1/j$. We set $F = \cup_{j=1}^{\infty} F_j$. Then F is measurable and $E \setminus F \subseteq E \setminus F_j$ for each $j \in \mathbb{N}$. Thus we have $m^*(E \setminus F) = 0$. So $W = E \setminus F$ is a Lebesgue null set and so is measurable. In conclusion, $E = F \cup W$ is Lebesgue measurable. $\qquad\square$

11.3 Approximation by Intervals

We conclude this chapter by showing that any Lebesgue measurable set can be approximated by the finite union of bounded, open intervals. In this discussion we use the notation

$$S \triangle T \equiv (S \setminus T) \cup (T \setminus S).$$

Theorem 11.11 *Let $E \subseteq \mathcal{L}$ have finite Lebesgue measure. Let $\epsilon > 0$. Then there are bounded open intervals $I_1, I_2, \ldots, I_k$ so that, if $U = \cup_{j=1}^{k} I_j$, then $m(E \triangle U) < \epsilon$.*

Proof: As we have seen before, there is a collection $\{I_j\}_{j=1}^{k}$ of bounded, open intervals which cover E and so that, if $\mathcal{I} = \cup_{j=1}^{k} I_j$, then $m(\mathcal{I}) \leq m(E) + \epsilon/2$. We also know that there is a compact set $K \subseteq E$ such that $m(E \setminus K) < \epsilon/2$.

The Heine-Borel theorem tells us that finitely many of the I_j, say $I_1, I_2, \ldots,$

I_m, cover K. If we set $L = \cup_{j=1}^{m} I_j$, then $K \subseteq L \subseteq \mathcal{I}$ and $K \subseteq E \subseteq \mathcal{I}$. Therefore

$$
\begin{aligned}
m(E \triangle L) &= m(E \setminus L) + m(L \setminus E) \\
&\leq m(E \setminus K) + m(\mathcal{I} \setminus E) \\
&< \epsilon.
\end{aligned}
$$

That completes the proof.   □

You can check for yourself that this last theorem is also true for closed intervals, or half-open intervals, or even for pairwise disjoint intervals.

Exercises

1. Prove that Theorem 11.11 is true for closed intervals.

2. Prove that Theorem 11.11 is true for pairwise disjoint, open intervals.

3. Prove Proposition 11.6.

4. Prove Corollary 11.9.

5. Let $\mathcal{O} \subseteq \mathbb{R}$ be any bounded, open set. Show explicitly and constructively that $\mathcal{O}$ can be approximated from the inside by compact sets.

6. Let $\mathcal{C}$ be the Cantor ternary set. Show explicitly and constructively that $\mathcal{C}$ can be approximated from the outside by open sets.

7. Provide the details of the proof of Corollary 11.8.

12

Different Methods of Convergence

12.1 Review of Convergence Techniques

Let $(X, \mathcal{X}, \mu)$ be a measure space. In this book we have treated four different types of convergence of a sequence of $\{f_j\}$ functions to a limit function f:

(i) pointwise convergence: For each $\epsilon > 0$ and each $x \in X$ there is a number $J > 0$ such that, if $j > J$, then $|f_j(x) - f(x)| < \epsilon$.

(ii) uniform convergence: For each $\epsilon > 0$ there is a number $J > 0$ such that, if $j > J$ and $x \in X$, then $|f_j(x) - f(x)| < \epsilon$.

(iii) convergence almost everywhere: There exists a set $E \subseteq X$ with $\mu(E) = 0$ so that, for every $\epsilon > 0$, and each $x \in X \setminus E$, there is a number $J > 0$ such that, for $j > J$, we have $|f_j(x) - f(x)| < \epsilon$.

(iv) convergence in $\mathbf{L^p}$, $1 \leq \mathbf{p} < \infty$: For each $\epsilon > 0$ there is a number $J > 0$ so that, if $j > J$, then

$$\|f_j - f\|_{L^p} = \int |f_j(x) - f(x)|^p \, d\mu(x)^{1/p} < \epsilon \, .$$

It is clear that uniform convergence implies pointwise convergence. Also pointwise convergence implies convergence almost everywhere. In the case of a finite measure space, uniform convergence also implies convergence in L^p. The reverse implications are false.

EXAMPLE 12.1 Let $f_j(x) = \chi_{[j,\infty)}$. Then the f_j converge to the identically 0 function pointwise and almost everywhere, but not in L^p for any $p \geq 1$. They do not converge uniformly.

Let $g_j(x) = \chi_{[1,1+1/j]}$. These functions converge almost everywhere and in L^p to the identically 0 function. They do not converge pointwise, and they do not converge uniformly.

Proposition 12.2 Let $(X, \mathcal{X}, \mu)$ be a measure space. Assume that $\mu(X) < +\infty$. Let $\{f_j\}$ be a sequence of L^p functions that converges uniformly on X to a limit function f. Then $f \in L^p$ and the sequence $\{f_j\}$ converges in L^p to f.

Proof: Let $\epsilon > 0$ and choose a $J > 0$ such that when $j > J$ and $x \in X$, $|f_j(x) - f(x)| < \epsilon$. Observe that, if $j > J$,

$$\|f_j - f\|_{L^p} = \left\{ \int |f_j(x) - f(x)|^p \, d\mu \right\}^{1/p}$$

$$\leq \left\{ \int \epsilon^p \, d\mu \right\}^{1/p}$$

$$= \epsilon \mu(X)^{1/p} . \qquad (12.2.1)$$

We conclude that $\{f_j\}$ converges in L^p to f. $\qquad\qquad \square$

Proposition 12.3 *Let $(X, \mathcal{X}, \mu)$ be a measure space. Let $1 \leq p < \infty$. Let $\{f_j\}$ be a sequence in L^p which converges pointwise almost everywhere to a measurable function f. If there is a $g \in L^p$ such that*

$$|f_j(x)| \leq g(x) \quad \forall x \in X , \quad \forall j \in \mathbb{N}, \qquad (12.3.1)$$

then $f \in L^p$ and $f_j \to f$ in L^p.

Proof: Because of inequality (12.3.1), we see that $|f(x)| \leq g(x)$ almost everywhere. Since $g \in L^p$, we conclude from Corollary 4.6 that $f \in L^p$.

Observe that

$$|f_j(x) - f(x)|^p \leq [2g(x)]^p , \quad \text{a.e.}$$

Since $\lim_{j \to \infty} |f_j(x) - f(x)|^p = 0$ a.e. and $2^p g^p$ belongs to L^1, the Lebesgue dominated convergence theorem tells us that

$$\lim_{j \to \infty} \int |f_j - f|^p \, d\mu = 0 .$$

As a result, $f_j \to f$ in L^p. $\qquad\qquad \square$

Corollary 12.4 *Let $(X, \mathcal{X}, \mu)$ be a measure space. Assume that $\mu(X) < +\infty$. Let $1 \leq p < \infty$. Let $\{f_j\}$ be a sequence in L^p which converges almost everywhere to a measurable function f. If there is a constant $K > 0$ such that*

$$|f_j(x)| \leq K \quad \forall x \in X , \quad \forall j \in \mathbb{N}, \qquad (12.4.1)$$

then f belongs to L^p and the sequence $\{f_j\}$ converges to f in L^p.

Proof: If $\mu(x) < +\infty$, then the constant functions belong to L^p. So the function $g(x) \equiv K$ is in L^p. Now apply the proposition. $\qquad\qquad \square$

One might suppose that L^p convergence implies almost everywhere convergence. But the next example shows that that is not the case.

EXAMPLE 12.5 Let $X = [0,1]$, $\mathcal{B}$ be the Borel sets, and μ be Lebesgue measure. Consider the intervals in $[0,1]$ with dyadic[1] endpoints. Order these intervals in decreasing order of size. Let f_j be the characteristic function of the jth interval.

Then it is clear that the f_j tend to $f \equiv 0$ in L^p norm. But, if x is any point of $[0,1]$, then there is a subsequence f_{j_k} that equals 1 at x and there is another subsequence f_{j_ℓ} that equals 0 at x. So we do *not* have pointwise convergence at *any point* of the interval $[0,1]$.

12.2 Convergence in Measure

In this section we treat a new concept of convergence which is analytically useful. And it is intuitively appealing.

Definition 12.6 Let $(X, \mathcal{X}, \mu)$ be a measure space. A sequence $\{f_j\}$ of measurable functions is said to *converge in measure* to a measurable function f precisely when

$$\lim_{j \to \infty} \mu(\{x \in \mathbb{R} : |f_j(x) - f(x)| \geq \alpha\}) = 0 \tag{12.6.1}$$

for each $\alpha > 0$.

The sequence $\{f_j\}$ is said to be *Cauchy in measure* when

$$\lim_{j,k \to \infty} \mu(\{x \in \mathbb{R} : |f_j(x) - f_k(x)| \geq \alpha\}) = 0 \tag{12.6.2}$$

for each $\alpha > 0$.

EXAMPLE 12.7 Let $f_j(x) = \chi_{[j,\infty)}$. Then the f_j do not converge in measure. Indeed they are not Cauchy in measure.

Let $g_j(x) = \chi_{[1,1+1/j]}$. Then the g_j converge in measure to the identically 0 function.

Proposition 12.8 Let $(X, \mathcal{X}, \mu)$ be a measure space. If the functions $\{f_j\}$ converge in L^p, $1 \leq p < \infty$, then the sequence converges in measure.

Proof: Let $\alpha > 0$. Set

$$E_j(\alpha) = \{x \in \mathbb{R} : |f_j(x) - f(x)| \geq \alpha\}.$$

Then

$$\int |f_j - f|^p \, d\mu \geq \int_{E_j(\alpha)} |f_j - f|^p \, d\mu \geq \alpha^p \cdot \mu(E_j(\alpha)).$$

---

[1] A point is dyadic if it has the form $j/2^k$.

We know that $\|f_j - f\|_{L^p} \to 0$. Since $\alpha > 0$, we may conclude that $\mu(E_j(\alpha)) \to 0$ as $j \to \infty$. $\qquad\square$

The reader may note that Example 12.5 shows that a sequence of functions can converge in measure while not converging pointwise at any point. However the following result of F. Riesz tends to ameliorate the situation.

Proposition 12.9 (F. Riesz) *Let $(X, \mathcal{X}, \mu)$ be a measure space. Suppose that $\{f_j\}$ is a sequence of measurable functions that is Cauchy in measure. Then there is a subsequence which converges almost everywhere and in measure to a measurable limit function f.*

Proof: Choose a subsequence $\{f_{j_k}\}$ so that $E_k = \{x \in \mathbb{R} : |f_{j_{k+1}}(x) - f_{j_k}(x)| \geq 2^{-k}\}$ satisfies $\mu(E_k) < 2^{-k}$. Set $F_k = \cup_{j=k}^{\infty} E_j$. Thus $F_k \in \mathcal{X}$ and $\mu(F_k) < 2^{-(k-1)}$.

If $\ell \geq m \geq n$ and $x \notin F_n$, then

$$|f_{j_\ell}(x) - f_{j_m}(x)| \leq |f_{j_\ell}(x) - f_{j_{\ell-1}}(x)| + \cdots + |f_{j_{m+1}}(x) - f_{j_m}(x)|$$

$$\leq \frac{1}{2^{\ell-1}} + \cdots + \frac{1}{2^m}$$

$$< \frac{1}{2^{m-1}} . \qquad\qquad (12.9.1)$$

Let $F = \cap_{n=1}^{\infty} F_n$, so that $F \in \mathcal{X}$ and $\mu(F) = 0$. From the above reasoning it follows that $\{f_{j_m}\}$ converges on $X \setminus F$. If we define

$$f(x) = \begin{cases} \lim_{m \to \infty} f_{j_m}(x) & \text{if} \quad x \notin F, \\ 0 & \text{if} \quad x \in F, \end{cases}$$

then $\{f_{j_m}\}$ converges almost everywhere to the measurable function f. Passing to the limit in (12.9.1) as $\ell \to \infty$, we conclude that, if $m \geq n$ and $x \notin F_n$, then

$$|f(x) - f_{j_m}| \leq \frac{1}{2^{m-1}} \leq \frac{1}{2^{n-1}} .$$

This proves that the sequence $\{f_{j_m}\}$ converges uniformly to f on the complement of each set F_n.

To see that $\{f_{j_m}\}$ converges in measure to f, let α, ϵ be positive real numbers and choose n so large that

$$\mu(F_n) < 2^{-(n-1)} < \min(\alpha, \epsilon) .$$

If $m \geq n$, then the above estimate shows that

$$\{x \in \mathbb{R} : |f(x) - f_{j_m}(x)| \geq \alpha\} \subseteq \{x \in \mathbb{R} : |f(x) - f_{j_m}(x)| > 2^{-(n-1)}\}$$
$$\subseteq F_n .$$

Thus

$$\mu\{x \in \mathbb{R} : |f(x) - f_{j_m}(x)| \geq \alpha\}) \leq \mu(F_n) < \epsilon$$

for all $m \geq n$. We conclude that $\{f_{j_m}\}$ converges in measure to f. □

Corollary 12.10 *Let $\{f_j\}$ be a sequence of measurable functions which is Cauchy in measure. Then there is a measurable function f to which the sequence converges in measure. This limit function is uniquely determined almost everywhere.*

Proof: We know that there is a subsequence $\{f_{j_k}\}$ that converges in measure to f. To see that the entire sequence converges in measure to f, notice that

$$|f(x) - f_j(x)| \leq |f(x) - f_{j_k}(x)| + |f_{j_k}(x) - f_j(x)|.$$

Thus

$$\{x \in \mathbb{R} : |f(x) - f_j(x)| \geq \alpha\} \subseteq \left\{x \in \mathbb{R} : |f(x) - f_{j_k}(x)| \geq \frac{\alpha}{2}\right\}$$
$$\bigcup \left\{x \in \mathbb{R} : |f_{j_k}(x) - f_j(x)| \geq \frac{\alpha}{2}\right\}.$$

The convergence in measure of $\{f_j\}$ follows from this relation.

Now suppose that the sequence $\{f_j\}$ converges in measure both to f and to g. Since

$$|f(x) - g(x)| \leq |f(x) - f_j(x)| + |f_j(x) - g(x)|,$$

it follows that

$$\{x \in \mathbb{R} : |f(x) - g(x)| \geq \alpha\} \subseteq \left\{x \in \mathbb{R} : |f(x) - f_j(x)| \geq \frac{\alpha}{2}\right\}$$
$$\bigcup \left\{x \in \mathbb{R} : |f_j(x) - g(x)| \geq \frac{\alpha}{2}\right\},$$

hence, passing to the limit,

$$\mu(\{x \in \mathbb{R} : |f(x) - g(x)| \geq \alpha\}) = 0$$

for all $\alpha > 0$. Taking $\alpha = 1/n$ for $n \in \mathbb{N}$, we conclude that $f = g$ a.e. □

Certainly convergence in L^p implies convergence in measure. The converse is not true, as the next example shows.

EXAMPLE 12.11 Let μ be Lebesgue measure on the σ-algebra $\mathcal{B}$. Let $f_j(x) = j \cdot \chi_{[j,j+1/j]}$. Then f_j converges in measure to the identically 0 function, but the f_j do not converge in L^p norm.

However, with an additional hypothesis, we can obtain a positive result.

Proposition 12.12 *Let $\{f_j\}$ be a sequence of functions in L^p which converges in measure to a function f. Let $g \in L^p$ satisfy*

$$|f_j(x)| \leq g(x) \quad \text{a.e.}$$

Then $f \in L^p$ and the sequence $\{f_j\}$ converges to f in L^p.

Proof: If $\{f_j\}$ does not converge in L^p to f, then there exists a subsequence $\{f_{j_k}\}$ and an $\epsilon > 0$ such that

$$\|f_{j_k} - f\|_{L^p} > \epsilon \quad \text{for} \;\; k \in \mathbb{N}. \tag{12.12.1}$$

Since $\{f_{j_k}\}$ is a subsequence of $\{f_j\}$, we see that it converges in measure to f. By Proposition 12.9, there is a subsequence $\{f_{j_{k_\ell}}\}$ which converges almost everywhere and in measure to a function h. From the uniqueness part of Corollary 12.10, it follows that $h = f$ a.e. Since $\{f_{j_{k_\ell}}\}$ converges almost everywhere to f and is dominated by g, Proposition 12.3 implies that $\|f_{j_{k_\ell}} - f\|_{L^p} \to 0$. But this contradicts (12.12.1). $\qquad\square$

Our concluding result for this chapter is quite striking, and ties together many of the ideas that we have introduced.

Theorem 12.13 (Vitali) *Let $(X, \mathcal{X}, \mu)$ be a measure space. Let $\{f_j\}$ be a sequence in L^p, $1 \leq p < \infty$. Then the following three conditions taken together are necessary and sufficient for the L^p convergence of $\{f_j\}$ to f.*

    (i) *$\{f_j\}$ converges to f in measure.*

    (ii) *For each $\epsilon > 0$, there is a set $E_\epsilon \in \mathcal{X}$ with $\mu(E_\epsilon) < +\infty$, such that, if $F \in \mathcal{X}$ and $F \cap E_\epsilon = \emptyset$, then*

$$\int_F |f_j|^p \, d\mu < \epsilon^p \quad \text{for all} \;\; j \in \mathbb{N}.$$

    (iii) *For each $\epsilon > 0$, there is a $\delta > 0$ so that, if $E \in \mathcal{X}$ and $\mu(E) < \delta$, then*

$$\int_E |f_j|^p \, d\mu < \epsilon^p \quad \text{for all} \;\; j \in \mathbb{N}.$$

Proof: We know that L^p convergence implies convergence in measure. The fact that L^p convergence of the $\{f_j\}$ implies **(ii)** and implies **(iii)** is straightforward and we leave the matter as an exercise for the reader.

We shall next prove that the three given conditions taken together imply that $\{f_j\}$ converges in the L^p topology to f. Let $\epsilon > 0$. Let E_ϵ be as in statement **(ii)** and let $F = X \setminus E_\epsilon$. If the Minkowski inequality is applied to

$$f_j - f_k = [(f_j - f_k)\chi_{E_\epsilon}] + [f_j\chi_F - f_k\chi_F],$$

then we have

$$\|f_j - f_k\|_{L^p} \leq \left\{ \int_{E_\epsilon} |f_j - f_k|^p \, d\mu \right\}^{1/p} + 2\epsilon$$

for $j, k \in \mathbb{N}$.

Now let $\alpha = \epsilon[\mu(E_\epsilon)]^{-1/p}$. Let

$$H_{jk} = \{x \in E_\epsilon : |f_j(x) - f_k(x)| \geq \alpha\}.$$

Because of **(i)**, there is a number K such that, if $j, k \geq K$, then $\mu(H_{jk}) < \delta$. Another application of Minkowski together with **(iii)** gives

$$\left\{ \int_{E_\epsilon} |f_j - f_k|^p \, d\mu \right\}^{1/p} \leq \left\{ \int_{E_\epsilon \setminus H_{jk}} |f_j - f_k|^p \, d\mu \right\}^{1/p}$$

$$+ \left\{ \int_{H_{jk}} |f_j|^p \, d\mu \right\}^{1/p} + \left\{ \int_{H_{jk}} |f_k|^p \, d\mu \right\}^{1/p}$$

$$\leq \alpha[\mu(E_\epsilon)]^{1/p} + \epsilon + \epsilon$$

$$= 3\epsilon,$$

for $j, k \geq K$.

Combining this last result with the first inequality in the proof, we find that the sequence $\{f_j\}$ is Cauchy in L^p and hence convergent in L^p. Since we already know that $\{f_j\}$ converges in measure to f, we may conclude from the uniqueness part of Corollary 12.10 that $\{f_j\}$ converges to f in L^p. □

Exercises

In each exercise here, the measure space is $\mathbb{R}$ with the σ-algebra consisting of the Borel sets, and the measure is Lebesgue measure.

1. Let $f_j(x) = j^{-1/p} \cdot \chi_{[0,j]}(x)$ for $j = 1, 2, \ldots$. Show that the sequence $\{f_j\}$ converges uniformly to the $\equiv 0$ function, but that the sequence does *not* converge in L^p.

2. Let $g_j(x) = j \cdot \chi_{[1/j, 2/j]}(x)$. Prove that the sequence $\{g_j\}$ converges pointwise at every point to the $\equiv 0$ function, but the sequence does *not* converge in L^p.

3. Show that both of the sequences in Exercises 1 and 2 converge in measure.

4. Let $h_j(x) = \chi_{[j,j+1]}(x)$. Show that the sequence $\{h_j\}$ converges at every point to the $\equiv 0$ function, but that it does not converge in measure.

5. Prove that the sequence in Exercise 2 shows that convergence in measure does *not* imply L^p convergence, even when the measure space is finite.

6. Examine the sequence in Example 12.5. Show that it has a subsequence which converges almost everywhere to the $\equiv 0$ function. Is there also a subsequence which converges at *every* point?

7. Suppose that each f_j is the characteristic function of a set E_j. And assume that the sequence $\{f_j\}$ converges in the L^p norm to a limit function f. Prove that f is a characteristic function.

8. Show that the Lebesgue dominated convergence theorem still holds if almost everywhere convergence is replaced by convergence in measure.

9. Let $(X, \mathcal{X}, \mu)$ be a finite measure space. If f is a measurable function, define
$$\mathcal{R}(f) = \int \frac{|f|}{1 + |f|}\, d\mu.$$
Prove that a sequence $\{f_j\}$ of measurable functions converges in measure to f if and only if $\mathcal{R}(f_j - f) \to 0$.

10. Suppose that $\varphi : \mathbb{R} \to \mathbb{R}$ is continuous. Further suppose that $\{f_j\}$ is a sequence of measurable functions that converges almost everywhere to a measurable function f. Then show that $\varphi \circ f_j$ converges almost everywhere to $\varphi \circ f$.

11. Let $(X, \mathcal{X}, \mu)$ be a finite measure space. Let $1 \leq p < \infty$. Let $\varphi : \mathbb{R} \to \mathbb{R}$ be continuous and satisfy this condition:

> There is a constant $K > 0$ such that $|\varphi(t)| \leq K|t|$ for $|t| \geq K$.

Then show that $\varphi \circ f$ belongs to L^p for each $f \in L^p$.

12. Suppose that $\{f_j\}$ converges to f in L^p on a finite measure space $(X, \mathcal{X}, \mu)$. Let φ be continous and satisfy the displayed condition in Exercise 11. Show that $\{\varphi \circ f_j\}$ converges to $\varphi \circ f$ in L^p.

13. Prove that L^p convergence implies parts **(ii)** and **(iii)** of Theorem 12.13.

13

Measure on a Product Space

13.1 Product Measures

Higher dimensional real analysis is much more complex and fascinating than analysis on the real line. Thus we are certainly interested in doing measure theory in $\mathbb{R}^2$, for instance. We could, if we wished, use the open sets in $\mathbb{R}^2$ to generate the σ-algebra of Borel sets and proceed from there. But it is natural to think of $\mathbb{R}^2 = \mathbb{R} \times \mathbb{R}$ and to wonder whether the 1-dimensional Lebesgue measure on each of the $\mathbb{R}$ factors can somehow be combined to produce a product measure on $\mathbb{R}^2$.

It turns out that the answer is "yes," and in fact the two approaches described in the last paragraph turn out to be essentially equivalent. That is the subject of this chapter. Note that the big theorems on this topic are those of Tonelli and Fubini.

Definition 13.1 Let $(X, \mathcal{X})$ and $(Y, \mathcal{Y})$ be measure spaces. Then a set of the form $A \times B$, with $A \in \mathcal{X}$ and $Y \in \mathcal{Y}$ is called a *measurable rectangle*, or sometimes simply a *rectangle* in $Z \equiv X \times Y$. See Figure 13.1. We denote the collection of all finite unions of rectangles by $\mathcal{Z}$.

As an exercise, you should verify that every element of $\mathcal{Z}$ can be written as a finite pairwise *disjoint* union of rectangles.

Lemma 13.2 *The collection $\mathcal{Z}$ is an algebra of subsets of Z.*

Proof: Clearly the finite union of sets in $\mathcal{Z}$ is also in $\mathcal{Z}$. It is also straightforward to check that the complement of a rectangle in Z is (not necessarily itself a rectangle but) also an element of $\mathcal{Z}$. By de Morgan's law, we see that the complement of any set in $\mathcal{Z}$ belongs to $\mathcal{Z}$. $\qquad\square$

Definition 13.3 If $(X, \mathcal{X})$ and $(Y, \mathcal{Y})$ are measure spaces, then $\mathcal{Z} = \mathcal{X} \times \mathcal{Y}$ denotes the σ-algebra of subsets of $Z = X \times Y$ generated by rectangles $A \times B$ with $A \in \mathcal{X}$ and $B \in \mathcal{Y}$. A set in $\mathcal{Z}$ is called a *$\mathcal{Z}$-measurable set*, or sometimes just a *measurable subset of Z*.

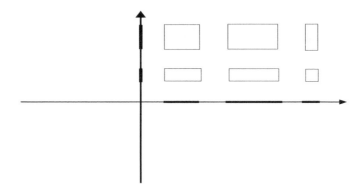

FIGURE 13.1
A measurable rectangle.

Theorem 13.4 (Product Measure Theorem) *Let $(X, \mathcal{X}, \mu)$ and $(Y, \mathcal{Y}, \nu)$ be measure spaces. Then there is a measure π defined on $\mathcal{Z} = \mathcal{X} \times \mathcal{Y}$ such that*

$$\pi(A \times B) = \mu(A) \cdot \nu(B) \qquad (13.4.1)$$

for all $A \in \mathcal{X}$ and $B \in \mathcal{Y}$. If the two measure spaces are σ-finite, then there is a unique measure π with property (13.4.1).

Proof: Suppose that the set $A \times B$ is the pairwise disjoint union of a collection $\{A_j \times B_j\}$ of rectangles, $j = 1, 2, \ldots$. Therefore

$$\chi_{A \times B}(x, y) = \chi_A(x) \cdot \chi_B(y) = \sum_{j=1}^{\infty} \chi_{A_j}(x) \cdot \chi_{B_j}(y)$$

for all $x \in X$, $y \in Y$. Holding x fixed, we integrate in y with respect to ν. Applying the monotone convergence theorem then yields

$$\chi_A(x) \cdot \nu(B) = \sum_{j=1}^{\infty} \chi_{A_j}(X) \cdot \nu(B_j).$$

A second application of the monotone convergence theorem then yields that

$$\mu(A) \cdot \nu(B) = \sum_{j=1}^{\infty} \mu(A_j) \cdot \nu(B_j).$$

Let $E \in \mathcal{Z}$. We may assume that

$$E = \bigcup_{j=1}^{\infty} (A_j \times B_j),$$

where the sets $A_j \times B_j$ are pairwise disjoint rectangles. If we define

$$\pi(E) = \sum_{j=1}^{\infty} \mu(A_j) \cdot \nu(B_j),$$

then the argument above implies that π is well defined and countably additive on $\mathcal{Z}$. By the Carathéodory extension theorem, there is an extension of π to a measure $\hat{\pi}$ on the σ-algebra $\hat{\mathcal{Z}}$ generated by $\mathcal{Z}$. Since $\hat{\pi}$ is an extension of π, it is clear that (13.4.1) holds.

In the case that $(X, \mathcal{X}, \mu)$ and $(Y, \mathcal{Y}, \nu)$ are σ-finite, then π is a σ-finite measure on the algebra $\hat{\mathcal{Z}}$ and the uniqueness of a measure satisfying (13.4.1) follows from the uniqueness assertion of the Hahn extension theorem. □

The measure constructed in Theorem 13.4 is called the *product* of μ and ν. In the case that μ and ν are both σ-finite, then they have a unique product.

Our next goal is, as we predicted in the introductory paragraph, to relate integration with respect to a product measure to iterated integration.

Definition 13.5 Let E be a subset of $Z = X \times Y$. Let $x \in X$. Then the *x-section* of E is the set

$$E_x = \{y \in Y : (x, y) \in E\}.$$

Similarly, if $y \in Y$, then the *y-section* of E is the set

$$E^y = \{x \in X : (x, y) \in E\}.$$

Definition 13.6 If $f : Z \to \hat{\mathbb{R}}$ and if $x \in X$, then the *x-section* of f is the function f_x defined on Y by

$$f_x(y) = f(x, y), \quad y \in Y.$$

Similarly, if $y \in Y$, then the *y-section* of f is the function f^y defined on X by

$$f^y(x) = f(x, y), \quad x \in X.$$

Lemma 13.7   (a) If E is a measurable subset of Z, then every section of E is measurable.

    (b) If $f : Z \to \hat{\mathbb{R}}$ is a measurable function, then every section of f is a measurable function.

Proof:

    (a) If $E = A \times B$ and $x \in X$, then the *x*-section of E is B if $x \in A$ and is $\emptyset$ if $x \notin A$. Thus the rectangles are contained in the collection $\mathcal{E}$ of all sets in $\mathcal{Z}$ having the property that each *x*-section is measurable. It is clear that $\mathcal{E}$ is a σ-algebra. Hence it follows that $\mathcal{E} = \mathcal{Z}$.

(b)   Let $x \in X$ and $\alpha \in \mathbb{R}$. Then

$$\{y \in Y : f_x(y) > \alpha\} \quad = \quad \{y \in Y : f(x,y) > \alpha\}$$
$$\leftrightarrow \quad \{(x,y) \in X \times Y : f(x,y) > \alpha\}_x\,.$$

Thus if f is $\mathcal{Z}$-measurable, then f_x is $\mathcal{Y}$-measurable. Similarly, f^y is $\mathcal{X}$-measurable.   □

Recall that a monotone class is a nonempty collection $\mathcal{M}$ of sets which contains the union of each increasing sequence in $\mathcal{M}$ and also the intersection of each decreasing sequence in $\mathcal{M}$. It is straightforward to check that if $\mathcal{A}$ is a nonempty collection of subsets of a set S, then the σ-algebra $\mathcal{S}$ generated by $\mathcal{A}$ contains the monotone class $\mathcal{M}$ generated by $\mathcal{A}$. Next we show that, if $\mathcal{A}$ is an algebra, then $\mathcal{S} = \mathcal{M}$.

Lemma 13.8 (Monotone Class Lemma) *If $\mathcal{A}$ is an algebra of sets, then the σ-algebra $\mathcal{S}$ generated by $\mathcal{A}$ coincides with the monotone class $\mathcal{M}$ generated by $\mathcal{A}$.*

Proof: Clearly $\mathcal{M} \subseteq \mathcal{S}$. To prove the reverse inequality it is sufficient (because $\mathcal{M}$ is also a monotone class) to prove that $\mathcal{M}$ is an algebra.

Now let $E \in \mathcal{M}$. Define $\mathcal{M}(E)$ to be the collection of sets $F \in \mathcal{M}$ so that $E \setminus F$, $E \cap F$, and $F \setminus E$ all belong to $\mathcal{M}$. Clearly $\emptyset, E \in \mathcal{M}(E)$. Also $\mathcal{M}(E)$ is obviously a monotone class. In addition, $F \in \mathcal{M}(E)$ if and only if $E \in \mathcal{M}(F)$.

If the set E belongs to the algebra $\mathcal{A}$, then clearly $\mathcal{A} \subseteq \mathcal{M}(E)$. But, since $\mathcal{M}$ is the smallest monotone class containing $\mathcal{A}$, we must have $\mathcal{M}(E) = \mathcal{M}$ for $E \in \mathcal{A}$. Thus, if $E \in \mathcal{A}$ and $F \in \mathcal{M}$, then $F \in \mathcal{M}(E)$. We conclude then that, if $E \in \mathcal{A}$ and $F \in \mathcal{M}$, then $E \in \mathcal{M}(F)$ and therefore $\mathcal{A} \subseteq \mathcal{M}(F)$ for any $F \in \mathcal{M}$. The minimality of $\mathcal{M}$ now implies that $\mathcal{M}(F) = \mathcal{M}$ for any $F \in \mathcal{M}$. So $\mathcal{M}$ is closed under intersection and relative complements. Since $X \in \mathcal{M}$, it is now plain that $\mathcal{M}$ is an algebra. Since $\mathcal{M}$ is also a monotone class, it is in fact a σ-algebra.   □

The monotone class lemma tells us in particular that, if a monotone class contains an algebra $\mathcal{A}$, then it contains the σ-algebra generated by $\mathcal{A}$.

Lemma 13.9 *Let $(X, \mathcal{X}, \mu)$ and $(Y, \mathcal{Y}, \nu)$ be σ-finite measure spaces. If $E \in \mathcal{Z} = \mathcal{X} \times \mathcal{Y}$, then the functions defined by*

$$f(x) = \nu(E_x) \quad \text{and} \quad g(y) = \mu(E^x) \tag{13.9.1}$$

are measurable, and

$$\int_X f \, d\mu = \pi(E) = \int_Y g \, d\nu\,. \tag{13.9.2}$$

Proof: Suppose at first that the two measure spaces are finite. Let $\mathcal{M}$ be the collection of all $E \in \mathcal{Z}$ for which (13.9.1) is true. We shall prove that $\mathcal{M} = \mathcal{Z}$ by showing that $\mathcal{M}$ is a monotone class containing the algebra $\mathcal{Z}$.

In point of fact, if $E = A \times B$ with $A \in \mathcal{X}$ and $B \in \mathcal{Y}$, then

$$f(x) = \chi_A(s) \cdot \nu(B) \quad , \quad g(y) = \chi_B(y) \cdot \mu(A) \, ,$$

are measurable and furthermore

$$\int_X f \, d\mu = \mu(A) \cdot \nu(B) = \int_Y g \, d\nu \, .$$

Since any element of $\mathcal{Z}$ can be written as a finite disjoint union of rectangles, we may conclude that $\mathcal{Z} \subseteq \mathcal{M}$.

Next we show that $\mathcal{M}$ is a monotone class. In fact, let $\{E_j\}$ be a monotone increasing sequence in $\mathcal{M}$ with union E. Then

$$f_j(x) = \nu((E_j)_x) \quad , \quad g_j(y) = \mu((E_j)^y)$$

are both measurable functions and

$$\int_X f_j \, d\mu = \pi(E_j) = \int_Y g_j \, d\nu \, .$$

Clearly the monotone increasing sequences $\{f_j\}$ and $\{g_j\}$ converge to the functions f and g defined by

$$f(x) = \nu(E_x) \quad \text{and} \quad g(y) = \mu(E^y)$$

respectively. If we use the fact that π is a measure together with the monotone convergence theorem, we may conclude that

$$\int_X f \, d\mu = \pi(E) = \int_Y g \, d\nu \, .$$

Hence $E \in \mathcal{M}$.

Since π is a finite measure, one can show in just the same way that if $\{F_j\}$ is a monotone decreasing sequence in $\mathcal{M}$, then $F = \cap_j F_j$ belongs to $\mathcal{M}$. Thus $\mathcal{M}$ is a monotone class, and it follows from the monotone class lemma that $\mathcal{M} = \mathcal{Z}$.

We refer the reader to [1] for the details of the σ-finite case. $\qquad\square$

Theorem 13.10 (Tonelli's Theorem) *Let $(X, \mathcal{X}, \mu)$ and $(Y, \mathcal{Y}, \nu)$ be σ-finite measure spaces. Let $Z = X \times Y$ and let $\psi : Z \to \widehat{\mathbb{R}}$ be measurable and nonnegative. Then the functions defined on X and Y by*

$$f(x) = \int_Y \psi_x \, d\nu \quad \text{and} \quad g(y) = \int_X \psi^y \, d\mu \qquad (13.10.1)$$

are measurable and

$$\int_X f \, d\mu = \int_Z \psi \, d\pi = \int_Y g \, d\nu \, . \qquad (13.10.2)$$

In other words,

$$\int_X \left(\int_Y \psi \, d\nu \right) d\mu = \int_Z \psi \, d\pi = \int_Y \left(\int_X \psi \, d\mu \right) d\nu . \qquad (13.10.3)$$

Proof: If ψ is the characteristic function of a set in Z, then the conclusion of the theorem is immediate from Lemma 13.9. By linearity, the theorem certainly holds for a measurable, simple function.

If now $\psi : Z \to \hat{\mathbb{R}}$ is an arbitrary nonnegative, measurable function, then we know that there is a sequence $\{s_j\}$ of nonegative, measurable simple functions which converge monotonically on Z to ψ. If φ_j, ψ_j are defined by

$$\varphi_j(x) = \int_Y (s_j)_x \, d\nu \quad \text{and} \quad \psi_j(x) = \int_X (s_j)^y \, d\mu , \qquad (13.10.4)$$

then φ_j, ψ_j are measurable and monotone in the index j. By the monotone convergence theorem, $\{\varphi_j\}$ converges on X to f and $\{\psi_j\}$ converges on Y to g.

Yet another application of the monotone convergence theorem yields

$$
\begin{aligned}
\int_X f \, d\mu &= \lim_{j\to\infty} \int_X \varphi_j \, d\mu \\
&= \lim_{j\to\infty} \int_Z s_j \, d\pi \\
&= \lim_{j\to\infty} \int_Y \psi_j \, d\nu \\
&= \int_Y g \, d\nu .
\end{aligned}
$$

In the same way one can show that

$$\int_Z \psi d\pi = \lim_{j\to\infty} \int_Z s_j d\pi .$$

From this we may conclude (13.10.2). $\qquad \square$

As we can see, Tonelli's theorem answers the main question posed at the start of this chapter about realizing integrals on product spaces in two different ways—but only for nonnegative functions. It was Fubini who realized how to treat the case of functions that take both positive and negative values.

Theorem 13.11 (Fubini's Theorem) *Let $(X, \mathcal{X}, \mu)$ and $(Y, \mathcal{Y}, \nu)$ be σ-finite measure spaces. Let the measure π on $\mathcal{Z} = \mathcal{X} \times \mathcal{Y}$ be the product of μ and ν. Let $Z = X \times Y$. If the function $\psi : Z \to \mathbb{R}$ is integrable with respect to π, then the extended real-valued functions defined almost everywhere by*

$$f(x) = \int_Y \psi_x \, d\nu \quad \text{and} \quad g(y) = \int_X \psi^y \, d\mu \qquad (13.11.1)$$

have finite integrals and

$$\int_X f \, d\mu = \int_Z \psi \, d\pi = \int_Y g \, d\nu \,. \tag{13.11.2}$$

In other words,

$$\int_X \left(\int_Y \psi \, d\nu \right) d\mu = \int_Z \psi \, d\pi = \int_Y \left(\int_X \psi \, d\mu \right) d\nu \,. \tag{13.11.3}$$

Proof: Since ψ is integrable with respect to π, then both its positive and negative parts ψ^+ and ψ^- are integrable. We apply Tonelli's theorem to ψ^+ and to ψ^- to conclude that the corresponding f^+ and f^- have finite integrals with respect to μ. Therefore f^+ and f^- are finite-valued μ-almost everywhere, so their difference f is defined μ-almost everywhere and the first part of (13.11.2) is clear. The second part of (13.11.2) is proved in a similar fashion. $\qquad\square$

It is worth noting that the functions defined in (13.11.1) are equal almost everywhere to integrable functions. The most important hypothesis in Fubini's theorem is that ψ be integrable.

Exercises

1. Give necessary and sufficient conditions on sets A and B so that $A \times B$ is empty.

2. Prove that if $A_1 \times B_1 = A_2 \times B_2 \neq \emptyset$, then $A_1 = A_2$ and $B_1 = B_2$.

3. Let $A \subseteq \mathbb{R}$ and $B \subseteq \mathbb{R}$ be closed intervals. Show that $(\mathbb{R} \times \mathbb{R}) \setminus (A \times B)$ is a finite union of rectangles. [**Hint:** You may find it helpful to draw a picture in the plane.]

4. Let $(X, \mathcal{X})$ and $(Y, \mathcal{Y})$ be measure spaces. Let $A_j \in \mathcal{X}$ and $B_j \in \mathcal{Y}$ for $j = 1, 2, \ldots, k$. Show that

$$\bigcup_{j=1}^{k} (A_j \times B_j)$$

can be written as the pairwise disjoint union of a finite number of rectangles in $X \times Y$.

5. Let $(X, \mathcal{B})$ be the measure space consisting of the real numbers equipped with the σ-algebra consisting of the Borel sets. Show that every open subset of $\mathbb{R} \times \mathbb{R}$ is an element of $\mathcal{B} \times \mathcal{B}$. Explain why $\mathcal{B} \times \mathcal{B}$ is the Borel algebra of $\mathbb{R} \times \mathbb{R}$.

6. Let $(X, \mathcal{X}, \mu)$ and $(Y, \mathcal{Y}, \nu)$ be measure spaces. Suppose that f is an $\mathcal{X}$-measurable function and g is a $\mathcal{Y}$-measurable function. Define $h(x, y) = f(x) \cdot g(y)$. Show that h is $\mathcal{X} \times \mathcal{Y}$-measurable.

7. Let E be a subset of $\mathbb{R}$. Define

$$\gamma(E) = \{(x, y) \in \mathbb{R} \times \mathbb{R} : x - y \in E\}.$$

If E is a Borel set in $\mathbb{R}$ (i.e., $E \in \mathcal{B}$), then show that $\gamma(E)$ is a Borel set in $\mathbb{R} \times \mathbb{R}$. Use this result to show that if $f : \mathbb{R} \to \mathbb{R}$ is a Borel measurable function, then the function $F(x, y) = f(x - y)$ is measurable with respect to $\mathcal{B} \times \mathcal{B}$.

8. Let E and F be subsets of $Z = X \times Y$. Let $x \in X$. Show that $(E \setminus F)_x = E_x \setminus F_x$. If $\{E_j\}$ are subsets of Z, then show that

$$\left(\bigcup_j E_j \right)_x = \bigcup (E_j)_x .$$

9. Use the theorems from this chapter to prove that, if $a_{j,k} \geq 0$ for $j, k \in \mathbb{N}$, then

$$\sum_{j=1}^{\infty} \sum_{k=1}^{\infty} a_{j,k} = \sum_{k=1}^{\infty} \sum_{j=1}^{\infty} a_{j,k} .$$

10. Let $a_{j,k}$ be defined for $j, k \in \mathbb{N}$ by the conditions $a_{j,j} = +1$, $a_{j,j+1} = -1$, and $a_{j,k} = 0$ if $j \neq k$ or $j \neq k + 1$. Show that

$$\sum_{j=1}^{\infty} \sum_{k=1}^{\infty} a_{j,k} = 0$$

while

$$\sum_{k=1}^{\infty} \sum_{j=1}^{\infty} a_{j,k} = 1 .$$

Explain why this shows that the hypothesis of integrability in Fubini's theorem is really needed.

11. Let f be integrable on $(X, \mathcal{X}, \mu)$ and g be integrable on $(Y, \mathcal{Y}, \nu)$. Let $Z = X \times Y$. Define h on $Z = X \times Y$ by $h(x, y) = f(x) \cdot g(y)$. If π is the product of μ and ν, then show that h is π integrable and

$$\int_Z h \, d\pi = \left(\int_X d\mu \right) \times \left(\int_Y g \, d\nu \right) .$$

12. Discuss the necessity of the hypothesis of σ-finite in both Tonelli's and Fubini's theorems.

14

Additivity for Outer Measure

This chapter presents the striking fact that the outer measure m^* is additive over the union of two disjoint sets provided that only *one of them* is measurable. We will also establish some other additivity and non-additivity properties of m^*.

Theorem 14.1 *Let E be a Lebesgue measurable subset of $\mathbb{R}$ and let F be any subset of $\mathbb{R}$. Then*

(a) $m^*(E \cup F) + m^*(E \cap F) = m(E) + m^*(F)$.

(b) *If $E \cap F = \emptyset$, then $m^*(E \cup F) = m(E) + m^*(F)$.*

(c) *If $m(E) < \infty$ and if $E \subseteq F$, then we have $m^*(F \setminus E) = m^*(F) - m(E)$.*

Proof: Since $E \in \mathcal{L}$, it follows from Carathéodory's condition that $m^*(A) = m^*(A \cap E) + m^*(A \setminus E)$ for any set $A \subseteq \mathbb{R}$. In fact let $A = E \cup F$. Then

$$
\begin{aligned}
m^*(E \cup F) &= m^*((E \cup F) \cap E) + m^*((E \cup F) \setminus E) \\
&= m(E) + m^*(F \setminus E).
\end{aligned}
$$

If we take $A = F$, then we have

$$
m^*(F) = m^*(F \cap E) + m^*(F \setminus E).
$$

Thus we have

$$
\begin{aligned}
m^*(E \cup F) + m^*(E \cap F) &= [m(E) + m^*(F \setminus E)] + m^*(E \cap F) \\
&= m(E) + [m^*(F \setminus E) + m^*(E \cap F)] \\
&= m(E) + m^*(F).
\end{aligned}
$$

This establishes **(a)**.

For **(b)**, if $E \cap F = \emptyset$, then $m^*(E \cap F) = 0$, thus the desired conclusion is immediate.

For **(c)**, let $G = F \setminus E$. Hence $F = E \cup G$ and $E \cap G = \emptyset$. From **(b)** we then conclude that

$$
\begin{aligned}
m^*(F) &= m^*(E \cup G) \\
&= m(E) + m^*(G) \\
&= m(E) + m^*(F \setminus E).
\end{aligned}
$$

Since $m(E) < \infty$, we have that $m^*(F)$ and $m^*(F \setminus E)$ are either both $+\infty$ or both finite. Thus **(c)** follows.   □

We saw earlier that, for any set $E \subseteq \mathbb{R}$, there is a G_δ set H so that $E \subseteq H$ and $m^*(E) = m(H)$. Furthermore, E is Lebesgue measurable if and only if $H \setminus E$ is a null set. The next result is in the same vein, but from the perspective of "within."

Theorem 14.2 *If $m^*(E) < \infty$, then E is measurable if and only if there is a measurable set $W \subseteq E$ with $m(W) = m^*(E)$.*

Proof: If E is measurable then clearly we can just take $W = E$.

For the converse, suppose that $W \in \mathcal{L}$, $W \subseteq E$, and $m(W) = m^*(E)$. It then follows from Theorem 14.1**(c)** that

$$m^*(E \setminus W) = m^*(E) - m(W) = 0.$$

As a result, the null set $E \setminus W$ is Lebesgue measurable and hence $E = (E \setminus W) \cup W$ is also measurable.   □

14.1   A New Look at Carathéodory

Now we present a refined and, in effect, simpler version of Carathéodory's criterion.

Theorem 14.3 *Let $A \subseteq \mathbb{R}$ be Lebesgue measurable with $m(A) < \infty$. Then $E \subseteq A$ is Lebesgue measurable if and only if*

$$m(A) = m^*(E) + m^*(A \setminus E). \tag{14.3.1}$$

Proof: If E is measurable, then the claimed result is immediate from Carathéodory's condition.

Conversely, by Theorem 11.2 applied to $A \setminus E$, there is a G_δ set W with $A \setminus E \subseteq W$ and $m^*(A \setminus E) = m(W)$. Since $A \setminus E \subseteq A \cap W \subseteq W$, we may infer that

$$m^*(A \setminus E) \le m(A \cap W) \le m(W) = m^*(A \setminus E).$$

Thus we conclude that $m(A \cap W) = m^*(A \setminus E)$. But $A \cap W$ is measurable. Also

$$A \cap (A \cap W) = A \cap W \quad \text{and} \quad A \setminus (A \cap W) = A \setminus W,$$

so we conclude that

$$m(A) = m(A \cap W) + m(A \setminus W) = m^*(A \setminus E) + m(A \setminus W).$$

Using now equation (14.3.1), we find that

$$m(A \setminus W) = m^*(E).$$

Let $B = A \setminus W \subseteq E$. It follows from Theorem 14.2 that E is Lebesgue measurable. $\qquad\square$

Often we are studying sets which all lie in a large interval of the form $I_n = [-n, n]$ for n a large positive integer. The following result is then useful.

Corollary 14.4 *A set $E \subseteq I_n$ is Lebesgue measurable if and only if*

$$m(I_n) = m^*(E) + m^*(I_n \setminus E).$$

Proof: This result is immediate from Theorem 14.3 and the fact that I_n is measurable. $\qquad\square$

When we are studying unbounded sets, the following result is often useful.

Theorem 14.5 *A set $E \subseteq \mathbb{R}$ is Lebesgue measurable if and only if the sets $E \cap I_n$ are measurable for each $n \in \mathbb{N}$.*

Proof: If E is measurable then the result is obvious.

Conversely, if each $E_n = E \cap I_n$ is measurable for each n, then it follows from the fact that $E = \cup_{n=1}^{\infty} E_n$ that E is measurable. $\qquad\square$

14.2   A Few Words about Inner Measure

The most obvious notion of inner measure would be to approximate a measurable set from the inside by intervals. But (see Exercise 1 below) this is unrealistic because there are many sets of positive measure that contain no intervals.

An alternative approach is this. Suppose that $E \subseteq I_n$ for some n. Define the *inner measure $m_*(E)$* of E to be

$$m_*(E) = m(I_n) - m^*(I_n \setminus E).$$

Now Corollary 14.4 takes this form:

Corollary 14.6 *A set $E \subseteq I_n$ is Lebesgue measurable if and only if its inner measure and its outer measure are equal.*

This last result is useful, for instance, if one is studying sets which are all subsets of a fixed bounded interval.

Exercises

1. Give an example of a measurable set of positive measure that contains no nontrivial interval. [**Hint:** Think in terms of a Cantor-type set.]

2. Prove Corollary 14.6.

3. Prove that a null set can always be written as the countable union of bounded null sets.

4. Give an example of a set of positive measure that can be written as the *uncountable* union of null sets.

5. Calculate explicitly the inner measure of the Cantor ternary set.

6. Calculate explicitly the inner measure of the unit interval $[0, 1]$.

7. Show explicitly that the inner measure and the outer measure of the Cantor ternary set are equal.

15

Nonmeasurable Sets And Non-Borel Sets

15.1 Nonmeasurable Sets

In this penultimate chapter we talk about nonmeasurable sets and non-Borel sets. Some interesting examples are generated along the way.

Definition 15.1 If $A \subseteq \mathbb{R}$, then its *difference set* is defined to be

$$A \ominus A = \{a - b : a \in A, b \in A\}.$$

Note that, if $A \subseteq B$, then $A \ominus A \subseteq B \ominus B$.

Lemma 15.2 *Let $K \subseteq \mathbb{R}$ be a compact set with $m(K) > 0$. Then the difference set $K \ominus K$ contains an open interval with center at the origin of coordinates.*

Proof: Since $0 < m(K) < \infty$, there exists an open set U with $K \subseteq U$ and $M(U) < 2m(K)$. Because K is compact and $^cU = \mathbb{R} \setminus U$ is closed, we have that

$$\delta = \text{dist}(K, {}^cU) > 0.$$

Thus we have that if $|x| = \text{dist}(x, 0) < \delta$, then $x + K \subseteq U$.

We claim that $(x + K) \cap K \neq \emptyset$. If not, then since $K \cup (x + K) \subseteq U$ and $(x + K) \cap K = \emptyset$ and m is additive,

$$
\begin{aligned}
2m(K) &= m(K) + m(x + K) \\
&= m(K \cup (x + K)) \\
&\leq M(U) < 2m(K).
\end{aligned}
$$

That is a contradiction.

We conclude then that $(x + K) \cap K \neq \emptyset$ for all x with $|x| < \delta$. But then we have that, if $|x| < \delta$, then there exist $k_1, k_2 \in K$ such that $x = k_1 - k_2 \in K \ominus K$. Thus the set $K \ominus K$ contains the open ball with center 0 and radius δ. $\square$

Theorem 15.3 *Let $E \subseteq \mathbb{R}^N$ be any Lebesgue measurable set having positive measure. Then the difference set $E \ominus E$ contains an open interval centered at 0.*

Proof: For $n \in \mathbb{N}$, let $E_n = \{x \in E : |x| < n\}$. Since $m(E) = \lim_{n \to \infty} m(E_n)$, we see that $m(E_n) > 0$ for n sufficiently large—say that $n \geq n_0$. Certainly $0 < m(E_{n_0}) < \infty$. Thus, by Theorem 11.10, there exists a compact set $K \subseteq E_{n_0} \subseteq E$ such that

$$0 < \frac{1}{2} \cdot m(E_{n_0}) \leq m(K).$$

Since $K \subseteq E$, we know that $K \ominus K \subseteq E \ominus E$. The preceding lemma now tells us that $K \ominus K$ contains an open interval with center 0. Hence so does $E \ominus E$. $\square$

Theorem 15.4 *A set $E \subseteq \mathbb{R}$ with positive outer measure contains a non-measurable subset.*

Proof: Review the argument presented in Section 1.2 which established the existence of a nonmeasurable set. We called that nonmeasurable set S, and we considered its translates $S_q = S + q = \{s + q : s \in S\}$ for each rational q. Of course each S_q is nonmeasurable, just because it is a translate of S. But it is conceivable that, for some q, $E_q \equiv S_q \cap E$ *is* measurable.

However, if E_q is measurable for some q and has positive measure, then Theorem 15.3 tells us that the difference set $E_q \ominus E_q$ must contain a nontrivial open interval. Since $E_q \subseteq S_q$, it follows that $S_q \ominus S_q$ also must contain a nontrivial open interval. That contradicts the construction of S. We conclude that those sets E_q which are measurable must be null sets.

Again referring to our argument in Section 1.2, we have that

$$E = \bigcup_{q \in \mathbb{Q}} E \cap S_q = \bigcup_{q \in \mathbb{Q}} E_q.$$

If all of the sets E_q are measurable, then we know that they must be null sets. Hence E is a null set, and that is a contradiction. Thus at least one of the sets E_q is nonmeasurable. $\square$

Now we strengthen this result as follows.

Theorem 15.5 *Let E be a Lebesgue measurable set such that $0 < m(E) < \infty$. Then there exist nonmeasurable subsets S and T of E such that $E = S \cup T$, $S \cap T = \emptyset$, and*

$$m(E) < m^*(S) + m^*(T). \tag{15.5.1}$$

Proof: Theorem 15.4 tells us that the set E has a nonmeasurable subset S. Set $T = E \setminus S$. Thus $E = S \cup T$ and $S \cap T = \emptyset$. Furthermore, since $S = E \setminus T$, the set T must also be nonmeasurable.

The subadditivity of m^* now tells us that

$$m^*(E) \leq m^*(S) + m^*(T). \tag{15.5.2}$$

If equality holds in (15.5.2), then Theorem 14.3 tells us that S is a measurable set, which is false. Thus we have (15.5.1). □

Next we shall establish that every nonmeasurable set with finite outer measure is part of a nonadditive decomposition of a measurable set.

Theorem 15.6 *Let S be a nonmeasurable set so that $m^*(S) < \infty$. Let H be a G_δ set with $S \subseteq H$ and $m^*(S) = m(H)$. Then $T = H \setminus S$ is also nonmeasurable and*

$$m(H) = m(S \cup T) < m^*(S) + m^*(T).$$

Proof: The existence of H was established in Lemma 11.2. Since $S = H \setminus T$, we see that T must be nonmeasurable. Thus $m^*(T) > 0$. The strict inequality then follows. □

15.2 Existence of a Measurable Set That Is Not Borel

Here we present a construction from [1, Ch. 17] of a Lebesgue measurable set that is not Borel. As part of this argument, we use the famous Cantor-Lebesgue function which you probably learned about in your undergraduate analysis course. We review that function here.

Let $\mathcal{C}$ be the Cantor ternary set. Each element x of the Cantor set has a ternary (instead of a decimal) expansion which can be written as

$$x = 0.c_1 c_2 c_3 \cdots,$$

where each c_j is either a 0 or a 2. Let $I = [0, 1]$ be the unit interval. We define a mapping $\varphi : \mathcal{C} \to I$ by

$$\varphi(x) = 0.(c_1/2)(c_2/2)(c_3/2) \cdots.$$

Here the number on the right should be interpreted as a binary expansion (all the digits are 0s or 1s).

Clearly, if $x, x^* \in \mathcal{C}$ and $x < x^*$, then there is a $j \in \mathbb{N}$ such that all the digits in the ternary expansion of x equal the corresponding digits in the ternary expansion of x^* up to the jth digit, but the jth digit of x is 0 while the jth digit of x^* is 2. It follows then that $\varphi(x) \leq \varphi(x^*)$ so that φ is a monotone nondecreasing function from $\mathcal{C}$ to I.

We should note in passing that φ is *not* one-to-one. For example, if $x = 0.020\underline{2}$ and $x^* = 0.022\underline{0}$ then certainly $x < x^*$. [Here we underline a digit if it is repeated infinitely often.] Yet $\varphi(x) = 0.010\underline{1} = 0.011\underline{0} = \varphi(x^*)$. In fact the reader may check that $\varphi(x) = \varphi(x^*)$ precisely when x, x^* are the left and right endpoints of one of the removed ternary intervals in the construction of the Cantor set.

It should be noted, however, that φ does map $\mathcal{C}$ *onto* I. This is so because if $y = 0.b_1b_2b_3 \cdots$ is the binary expansion of any point y in I, then y is the image under φ of the point x which has ternary expansion $0.(2b_1)(2b_2)(2b_3) \cdots$.

Next we extend φ to be defined on all of I by defining it to be constant on each of the removed ternary intervals. In fact φ takes the same value at both endpoints of such a removed ternary interval, so that tells us what the constant value should be on that interval. As an instance,

$$\varphi(x) = 0.0\underline{1} = 0.1\underline{0} = \frac{1}{2}$$

for all x satisfying $0.02 = 1/3 \le x \le 2/3 = 0.20$ in ternary.

This extended function, which we continue to denote by φ, is plainly a monotone nondecreasing function mapping I into I. It does not have any jump discontinuities (which are the only kind of discontinuities that a monotone function can have) since every value in I is taken at least once. So the extended function φ is continuous at every point of I.

It is worth noting that the derivative φ' equals 0 at each point of $I \setminus \mathcal{C}$, since φ is constant on a neighborhood of such a point. The extended function φ is usually called the *Cantor-Lebesgue function*. It is worth noting explicitly that $\varphi'(x) = 0$ for almost every $x \in I$.

Now we define

$$\psi(x) = \varphi(x) + x \,.$$

This new function ψ is a strictly increasing function from I to the closed interval $[0, 2]$. So it is one-to-one. And it is onto. It is also continuous. Hence the inverse function from $[0, 2]$ to I is also continuous. In other words, ψ is a homeomorphism from I to $[0, 2]$. It follows that both ψ and ψ^{-1} take Borel sets to Borel sets.

Since φ is constant on each of the ternary intervals that was removed in the construction of the Cantor set, we see that ψ maps each such interval into an interval of the same length. Thus

$$m(\psi(I \setminus \mathcal{C})) = m(I \setminus \mathcal{C}) = 1 \,.$$

Since $m([0, 2]) = 2$ and $[0, 2] = \psi(\mathcal{C}) \cup \psi(I \setminus \mathcal{C})$ and $\psi(\mathcal{C}) \cap \psi(I \setminus \mathcal{C}) = \emptyset$, we see that

$$2 = m(\psi(\mathcal{C})) + m(\psi(I \setminus \mathcal{C})) \,.$$

As a result, $m(\psi(\mathcal{C})) = 1$. In conclusion, the homeomorphism ψ maps the set $\mathcal{C}$, which has Lebesgue measure 0, to a set with Lebesgue measure 1.

Since $\psi(\mathcal{C})$ has positive measure, we know from Theorem 15.4 that it contains a set V that is *not* Lebesgue measurable. Then the set $V^* = \psi^{-1}(V)$ is a subset of $\mathcal{C}$ and hence is a Lebesgue null set. Therefore V^* *is* Lebesgue measurable. However V^* cannot be a Borel set; if it were, then $V = \psi(V^*)$

would also be Borel, and hence Lebesgue measurable. But this contradicts the choice of V as a nonmeasurable set.

We summarize the point of our analysis with an enunciated theorem.

Theorem 15.7 *There exists a Lebesgue measurable subset of $\mathbb{R}$ that is not Borel.*

One upshot of the proof that we just presented is that it is possible for a homeomorphism to map a Lebesgue measurable set to a nonmeasurable set.

Exercises

1. From our discussions in Chapter 13, you are familiar with the idea of product measure. Using those ideas, give an example of a set in the plane which is Lebesgue measurable but not Borel.

2. Let $\mathcal{C}$ be the Cantor ternary set. Calculate $\mathcal{C} \ominus \mathcal{C}$.

3. Let $\mathbb{Q}$ be the rational numbers. Calculate $\mathbb{Q} \ominus \mathbb{Q}$. What happens if you replace the rational numbers by the irrational numbers?

4. Refer to Section 1.2 for the construction of a nonmeasurable set. What can you say about the outer measure of this set?

5. Let S be the nonmeasurable set from Section 1.2. What can you say about $S \ominus S$?

6. In our first argument that there exists a measurable set that is not Borel, we showed that the cardinality of the measurable sets exceeds the cardinality of the Borel sets. How does the cardinality of the nonmeasurable sets compare to the cardinality of the measurable sets?

7. We know that any set of positive outer measure contains a non-measurable set. But this assertion is not true for null sets. Why not?

8. Let S be a nonmeasurable set and let E be a null set. What can you say about the measurability of $S \cup E$?

16

Applications

We take this opportunity to provide some elegant applications of Lebesgue measure theory to basic harmonic analysis. This will give the reader a glimpse of how useful Lebesgue measure can be.

Definition 16.1 Let f be an integrable function on $\mathbb{R}$. Define the *Hardy-Littlewood maximal function* Mf of f to be

$$Mf(x) = \sup_{R>0} \frac{1}{2R} \int_{x-R}^{x+R} |f(t)| \, d\mu(t).$$

In the twentieth century, the Hardy-Littlewood maximal function became a fundamental tool of analysis. We shall learn two nice uses of this concept.

Definition 16.2 Let f be a measurable function on $\mathbb{R}$. We say that f is *weak type 1* if there is a constant $C > 0$ so that, for each $\lambda > 0$,

$$\mu\{x \in \mathbb{R} : |f(x)| > \lambda\} \leq \frac{C}{\lambda}.$$

EXAMPLE 16.3 The function $f(x) = 1/x$ is weak type 1 on $\mathbb{R}$. It is *not*, however, Lebesgue integrable.

Definition 16.4 Let T be a linear operator on the space $L^1(\mathbb{R})$ of Lebesgue integrable functions, taking values in the measurable functions. We say that T is of *weak type* $(1,1)$ if there is a positive constant C so that, for each $f \in L^1$ and each $\lambda > 0$,

$$\mu\{x \in \mathbb{R} : |Tf(x)| > \lambda\} \leq \frac{C\|f\|_{L^1}}{\lambda}.$$

Theorem 16.5 *The Hardy-Littlewood maximal operator M is weak type* $(1,1)$.

In order to prove this result we need a geometric result that is commonly known as a *covering lemma*.

Lemma 16.6 *Let $K \subseteq \mathbb{R}$ be a compact set. Let $\{I_j\}$ be a finite collection of open intervals that covers K in the sense that $K \subseteq \cup_j I_j$. Write $I_j = (c_j - r_j, c_j + r_j)$, each j. Then there is a pairwise disjoint subcollection $\{I_{j_k}\}$ with the property that the three-fold dilates $3I_j = (c_j - 3r_j, c_j + 3r_j)$ cover K.*

Proof: Choose I_{j_1} which has the greatest length. If there is more than one then simply select one. Next choose I_{j_2} which is disjoint from I_{j_1} and which has greatest length. If there is more than one then simply select one. Continue in this fashion until the process stops—and it must stop because the collection $\{I_j\}$ is finite.

Now the chosen $\{I_{j_k}\}$ is a pairwise disjoint collection and we claim that $\{3I_{j_k}\}$ covers K. It suffices to show that $\{3I_{j_k}\}$ covers each of the original intervals I_j.

So fix an I_j. Let I_{j_k} be the first of the selected intervals that intersects I_j. Then the length of I_{j_k} must be greater than or equal to that of I_j by the way that we selected the I_{j_k}. But then it follows from the triangle inequality that $3I_{j_k}$ contains I_j. And that is what we wished to prove. □

Proof of Theorem 16.5: Fix $f \in L^1(\mathbb{R})$. Fix $\lambda > 0$. Let

$$S_\lambda = \{x \in \mathbb{R} : Mf(x) > \lambda\} \ .$$

Let K be a compact subset of S_λ. We shall prove that

$$\mu(K) \leq C \frac{\|f\|_{L^1}}{\lambda} \ .$$

The inner regularity of the Lebesgue measure will then yield the desired result.

Note that, for each $k \in K$, there is an interval $(k - \epsilon_k, k + \epsilon_k)$ so that

$$\frac{1}{2\epsilon_k} \int_{k-\epsilon_k}^{k+\epsilon_k} |f(t)| \, d\mu(t) > \lambda \ .$$

Now the intervals $I_k = (k-\epsilon_k, k+\epsilon_k)$ cover K. We may use the compactness of K to pass to a finite subcover. Then we may invoke the covering lemma to find a pairwise disjoint subcollection $\{I_{k_\ell}\}_{\ell=1}^m$ whose threefold dilates still cover K. Then we have

$$
\begin{aligned}
\mu(K) \ &\leq \ \mu\left(\bigcup_{\ell=1}^m 3I_{k_\ell}\right) \\
&\leq \ \sum_{\ell=1}^m \mu(3I_{k_\ell}) \\
&= \ \sum_{\ell=1}^m 3\mu(I_{k_\ell}) \\
&\leq \ 3\sum_{\ell=1}^m \frac{1}{\lambda} \int_{I_{k_\ell}} |f(t)| \, d\mu(t) \\
&\leq \ \frac{3}{\lambda} \cdot \|f\|_{L^1} \ .
\end{aligned}
$$

In the last inequality we have of course used the fact that the I_{k_ℓ} are pairwise disjoint.

That proves the result. □

Now we formulate and prove a version of the celebrated Lebesgue differentiation theorem.

Theorem 16.7 *Let f be a Lebesgue integrable function on $\mathbb{R}$. Then, for almost every $x \in \mathbb{R}$, we have that*

$$\lim_{R \to 0} \frac{1}{2R} \int_{-R}^{R} f(t)\, dt = f(x).$$

Proof: Fix $f \in L^1(\mathbb{R})$. Let $\epsilon > 0$. Choose a continuous, compactly supported function g so that $\|f - g\|_{L^1} < \epsilon^2$. Then we have

$$\mu\left\{x \in \mathbb{R} : \left|\limsup_{R \to 0} \frac{1}{2R} \int_{-R}^{R} f(t)\, dt - \liminf_{R \to 0} \frac{1}{2R} \int_{-R}^{R} f(t)\, dt\right| > \epsilon\right\}$$

$$\leq \quad \mu\{x \in \mathbb{R} : \limsup_{R \to 0} \frac{1}{2R} \int_{-R}^{R} |f(t) - g(t)|\, dt > \epsilon/3\}$$

$$+ \mu\left\{x \in \mathbb{R} : \left|\limsup_{R \to 0} \frac{1}{2R} \int_{-R}^{R} g(t)\, dt - \liminf_{R \to 0} \frac{1}{2R} \int_{-R}^{R} g(t)\, dt\right| > \epsilon/3\right\}$$

$$+ \mu\{x \in \mathbb{R} : \limsup_{R \to 0} \frac{1}{2R} \int_{-R}^{R} |g(t) - f(t)|\, dt > \epsilon/3\}.$$

Now the middle term on the right is obviously 0 by the continuity of g. The first and last terms can be majorized as follows:

$$\leq \mu\{x \in \mathbb{R} : M(f - g) > \epsilon/3\} + \mu\{x \in \mathbb{R} : M(g - f) > \epsilon/3\}.$$

This in turn is

$$\leq 3 \cdot \frac{\|f - g\|_{L^1}}{\epsilon/3} + 3 \cdot \frac{\|f - f\|_{L^1}}{\epsilon/3} \leq 3\frac{\epsilon^2}{\epsilon/3} + 3\frac{\epsilon^2}{\epsilon/3} = 18\epsilon.$$

This proves that

$$\mu\left\{x \in \mathbb{R} : |\limsup_{R \to 0} \frac{1}{2R} \int_{-R}^{R} f(t)\, dt - \liminf_{R \to 0} \frac{1}{2R} \int_{-R}^{R} f(t)\, dt| > \epsilon\right\} = 0.$$

We have thus shown that the desired limit exists almost everywhere. Since, when f is continuous, it is obvious that the limit equals $f(x)$, it follows then that the limit equals f almost everywhere for any $f \in L^1$. □

We conclude with a useful result about averages of functions against dilates of testing functions.

Let φ be a continuous function with compact support. Assume that $0 \leq \varphi \leq 1$ and that φ vanishes outside the interval $[-1, 1]$. Also suppose that $\int \varphi(x)\, dx = 1$. For $\epsilon > 0$ define $\varphi_\epsilon(x) = \epsilon^{-1}\varphi(x/\epsilon)$. Then we have the following result.

Theorem 16.8 *Let f be a Lebesgue integrable function on $\mathbb{R}$. Then, for almost every $x \in \mathbb{R}$,*

$$\lim_{\epsilon \to 0} f * \varphi_\epsilon(x) = f(x).$$

Proof: We observe that

$$\varphi(x) \leq \frac{1}{1}\chi_{[-1,1]}.$$

As a result,

$$\varphi_\epsilon(x) \leq \epsilon^{-1}\chi_{[-\epsilon,\epsilon]}.$$

Because of this last estimate, we see that the proof of this new theorem is just as in the proof of the last theorem. We estimate the limsup minus the liminf with three terms, and we then estimate each of those three terms with the corresponding term that comes from the Hardy-Littlewood maximal function. We leave the details for the interested reader.   □

Exercises

1. Prove this strengthened version of the Lebesgue differentiation theorem:

   Theorem: Let $f \in L^1(\mathbb{R})$. Then, for almost every $x \in \mathbb{R}$,

   $$\lim_{R \to 0} \frac{1}{2R}\int_{-R}^{R} |f(t) - f(x)|\, d\mu(t) = 0.$$

   Why do we call this "strengthened"?

2. Define $\varphi(x) = (1/\sqrt{\pi})e^{-x^2}$. Let $\varphi_\epsilon(x) = \epsilon^{-1}\varphi(x/\epsilon)$. Show that, if $f \in L^1$, then

   $$\lim_{\epsilon \to 0} f * \varphi_\epsilon(x) = f(x)$$

   for almost every x.

3. Explain why we think of φ_ϵ in Theorem 16.8 as an approximation to the identity.

4. State and prove a version of Theorem 16.8 for $f \in L^p$. What is the correct range of p?

5. A function f is said to be *locally integrable* if it is integrable on compact sets. Give an example of a function that is locally integrable but not integrable.

6. Refer to Exercise 5 for terminology. Prove a version of Theorem 16.7 for f a locally integrable function.

7. Refer to Exercise 5 for terminology. Prove a version of Theorem 16.8 for f a locally integrable function.

Table of Notation

Notation	Section	Meaning
$[a, b]$	1.1	closed interval
$\mathcal{P}$	1.1	partition
I_j	1.1	jth interval in partition
$\triangle_j$	1.1	length of jth interval
$m(\mathcal{P})$	1.1	mesh of partition
$\mathcal{R}_\mathcal{P}$	1.1	Riemann sum
$\int_a^b f(x)\,dx$	1.1	Riemann integral
$m(S)$	1.2	length of the set S
σ-algebra	1.2	an algebra of sets satisfying subadditivity
$^c S$	1.3	the complement of the set S
$(X, \mathcal{X}, \mu)$	1.3	a measure space
$\mathcal{X}$	1.3	a σ-algebra
$\mathcal{B}$	1.3	the Borel sets
f^+	1.3	positive part of f
f^-	1.3	negative part of f
$\widehat{\mathbb{R}}$	1.3	the extended real number system
$+\infty,\ -\infty$	1.3	elements of the extended reals
$s(x)$	1.3	a simple function
$\mathbb{R}^+$	2.1	the set of nonnegative reals
$\widehat{\mathbb{R}}^+$	2.1	the nonnegative reals union $\{+\infty\}$
μ	2.1	a measure
a.e.	2.1	almost everywhere
μ-a.e.	2.1	μ almost everywhere
$\int_X f\,d\mu$	3.1	the Lebesgue integral
λ	4.1	a signed measure
V	5.1	a vector space
$\mathbf{u}, \mathbf{v}, \mathbf{w}$	5.1	vectors

Notation	Section	Meaning
N	5.1	a norm
$\|\ \|$	5.1	a norm
L^1	5.1	the space of integrable functions
L^p	5.1	the space of pth power integrable functions
$\|f\|_{L^1}$	5.1	the L^1 norm
$\|f\|_{L^p}$	5.1	the L^p norm
$\|f\|_{L^\infty}$	5.2	the L^∞ norm
P	7.1	a positive set
N	7.1	a negative set
$\mathcal{P}$	7.1	the collection of all positive sets
$\mathcal{N}$	7.1	the collection of all negative sets
λ^+	7.1	the positive variation of λ
λ^-	7.1	the negative variation of λ
$\lambda \ll \mu$	7.2	λ is absolutely continuous with respect to μ
$\lambda \perp \mu$	7.2	λ and μ are mutually singular
φ^+	7.3	the positive part of the linear functional φ
φ^-	7.3	the negative part of the linear functional φ
$\mathcal{A}$	8.1	an algebra or field
$\widehat{\mu}$	8.1	outer measure
$\ell(S)$	8.1	length of S
$\widehat{\mathcal{A}}$	8.1	the σ-algebra containing $\mathcal{A}$
$\mathcal{F}^*$	8.2	the Lebesgue measurable sets
ℓ^*	8.2	Lebesgue measure
μ_g	8.3	Lebesgue-Stieltjes measure
γ	8.4	measure representing a linear functional on $C(X)$
$A \times B$	9.1	measurable rectangle
$\mathcal{Z}$	9.1	collection of all finite unions of rectangles
π	9.1	product measure
E_x	9.1	x-section of E
E^y	9.1	y-section of E
f_x	9.1	x-section of f
f^y	9.1	y-section of f
$\mathcal{M}$	9.1	monotone class
m^*	10.1	Lebesgue outer measure
E_a	10.1	the translate of E by a
$\mathcal{X}$	11.1	a σ-algebra
$\mathcal{L}$	11.1	the σ-algebra of Lebesgue measurable sets

Notation	Section	Meaning
m	11.1	Lebesgue measure
G_δ	12.1	countable intersection of open sets
F_σ	12.1	countable union of closed sets
$G_{\delta\sigma}$	12.1	countable union of G_δ sets
$F_{\sigma\delta}$	12.1	countable intersection of F_σ sets
$\aleph_0$	12.3	the cardinality of the natural numbers
c	12.3	the cardinality of the reals
$\mathcal{O}$	13.1	an open set
K	13.2	a compact set
I, I_j	13.3	open intervals
I_n	14.1	the interval $[-n, n]$
m_*	14.2	inner measure
$A \ominus A$	15.1	the difference set
$\mathcal{C}$	15.2	the Cantor ternary set
φ	15.2	the Cantor-Lebesgue function

Glossary

absolutely continuous   We say that the measure λ is absolutely continuous with respect to the measure μ if, whenever $E \in \mathcal{X}$ and $\mu(E) = 0$, then $\lambda(E) = 0$. We then write $\lambda \ll \mu$.

algebra   Let X be a given set. A family $\mathcal{A}$ of subsets of X is called an algebra if

   (i) $\emptyset$, X both belong to $\mathcal{A}$;
  (ii)   If E belongs to $\mathcal{A}$, then its complement $X \setminus E$ also belongs to $\mathcal{A}$;
 (iii)   If $E_1, E_2, \ldots, E_k$ belong to $\mathcal{A}$, then also their union $\cup_{j=1}^{k} E_j$ belongs to $\mathcal{A}$.

almost everywhere   A property holds almost everywhere if it holds at all points except on a set of measure zero.

almost everywhere convergence   A sequence $\{f_j\}$ of functions converges almost everywhere to f if there is a null set N such that $\lim_{j \to \infty} f_j(x)$ exists for all x not in N.

Axiom of Choice   The axiom of set theory that says that, if S is a set then there is a function that assigns to each subset T of S an element of T.

Banach space   A normed linear space that is complete.

Borel sets   The σ-algebra generated by the open sets.

Borel-Stieltjes measure   Let g be a monotone increasing function. The Borel-Stieltjes measure μ_g is a measure based on a notion of length for intervals where the length of an interval $[a, b]$ is $g(a) - g(b)$.

bounded linear functional   The linear functional φ is bounded if there is a constant $M > 0$ such that

$$|\varphi(f)| \leq M \|f\|_V$$

for all $f \in V$.

Cantor ternary set   A set originally constructed by Georg Cantor that is totally disconnected, has no interior, has zero length, but is uncountable.

Cantor-Lebesgue function   A function modeled on the Cantor set.

Carathéodory condition   Let m^* be the outer measure defined on all subsets of $\mathbb{R}$. A set $E \subseteq \mathbb{R}$ is said to satisfy the Carathéodory condition in case

$$m^*(A) = m^*(A \cap E) + m^*(A \setminus E) = m^*(A \cap E) + m^*(A \cap {}^c E)$$

for all subsets $A \subseteq \mathbb{R}$. The collection of all such sets will be denoted by $\mathcal{L}$.

Carathéodory Extension Theorem   The collection $\widehat{\mathcal{A}}$ of all $\widehat{\mu}$-measurable sets is a σ-algebra containing $\mathcal{A}$. Moreover, if $\{E_j\}$ is a pairwise disjoint sequence in $\widehat{\mathcal{A}}$, then

$$\widehat{\mu}\left(\bigcup_{j=1}^{\infty} E_j \right) = \sum_{j=1}^{\infty} \widehat{\mu}(E_j).$$

Carathéodory's theorem   Let m^* be the Lebesgue outer measure. The set $\mathcal{L}$ of all subsets of $\mathbb{R}$ that satisfy the Carathéodory condition is a σ-algebra of subsets of $\mathbb{R}$. Furthermore, the restriction of m^* to $\mathcal{L}$ is a measure on $\mathcal{L}$.

cardinality   A device due to G. Cantor for measuring the size of an infinite set.

Cauchy-Schwarz-Bunyakovskii inequality   The inequality

$$\left| \int fg \, d\mu \right| \leq \|f\|_{L^2} \cdot \|g\|_{L^2}.$$

Cauchy sequence   A sequence $\{f_j\}$ is said to be Cauchy if there is a number $J > 0$ so that if $j, k > J$, then $\|f_j - f_k\| < \epsilon$.

Cauchy sequence in measure   This is the condition that

$$\lim_{j,k \to \infty} \mu(\{x \in \mathbb{R} : |f_j(x) - f_k(x)| \geq \alpha\}) = 0 \tag{16.1}$$

for each $\alpha > 0$.

characteristic function   If S is a set then the associated characteristic function is

$$\chi_S(x) = \begin{cases} 1 & \text{if} & x \in S, \\ 0 & \text{if} & x \notin S. \end{cases}$$

complete measure   A measure with the property that any subset of a null set is measurable and has 0 measure.

complete space   A space V is complete if every Cauchy sequence in V converges to an element of V.

convergence in L^p   For each $\epsilon > 0$ there is a number $J > 0$ so that, if $j > J$, then

$$\|f_j - f\|_{L^p} = \int |f_j(x) - f(x)|^p \, d\mu(x)^{1/p} < \epsilon.$$

convergence in measure   A sequence $\{f_j\}$ of measurable functions is said to converge in measure to a measurable function f precisely when

$$\lim_{j \to \infty} \mu(\{x \in \mathbb{R} : |f_j(x) - f(x)| \geq \alpha\}) = 0 \qquad (16.2)$$

for each $\alpha > 0$.

convergent sequence   The sequence $\{f_j\}$ is said to be convergent to $f \in L^p$ if, for every $\epsilon > 0$, there is a number $J > 0$ so that if $j > J$, then $\|f_j - f\| < \epsilon$.

countable set   A set with the same cardinality as the set of natural numbers.

countable subadditivity   A set function μ is countably subadditive if

$$\mu \left(\bigcup_{j=1}^{\infty} E_j \right) \leq \sum_{j=1}^{\infty} \mu(E_j).$$

de Morgan's laws   Certain rules of logic that relate the negation of a conjunction to a disjunction and the negation of a disjunction to a conjunction.

denumerable set   A set which is either empty or finite or countable.

difference set   If $A \subseteq \mathbb{R}$, then its *difference set* is defined to be

$$A \ominus A = \{a - b : a \in A, b \in A\}.$$

differentiation under the integral sign   This is the performance of the following limit operation:

$$\frac{d}{dt}\int f(x,t)\,dx = \int \frac{d}{dt}f(x,t)\,dx\,.$$

essential lower bound   Let f be a measurable function. A real number a is called an essential lower bound for f if the set $f^{-1}((-\infty, a))$ has measure zero.

essential upper bound   Let f be a measurable function. A real number a is called an essential upper bound for f if the set $f^{-1}((a, \infty))$ has measure zero.

essentially bounded   The function f is essentially bounded if it has an essential upper bound and an essential lower bound.

extended Borel sets   The Borel sets with $\pm\infty$ allowed as members of a set.

extended real numbers   The real number system with $\pm\infty$ adjoined.

Fatou's lemma   Let $(X, \mathcal{X}, \mu)$ be a measure space. Assume that the functions f_j are nonnegative and measurable. Then

$$\int (\liminf_{j\to\infty} f_j)\,d\mu \le \liminf_{j\to\infty} \int f_j\,d\mu\,.$$

field   Let X be a given set. A family $\mathcal{A}$ of subsets of X is called a field if

(i) $\emptyset$, X both belong to $\mathcal{A}$;
(ii) If E belongs to $\mathcal{A}$, then its complement $X \setminus E$ also belongs to $\mathcal{A}$;
(iii) If $E_1, E_2, \ldots, E_k$ belong to $\mathcal{A}$, then also their union $\cup_{j=1}^{k} E_j$ belongs to $\mathcal{A}$.

finite measure   A measure which assigns a finite size or length to the entire space X.

Fubini's Theorem   Let $(X, \mathcal{X}, \mu)$ and $(Y, \mathcal{Y}, \nu)$ be σ-finite measure spaces. Let the measure π on $\mathcal{Z} = \mathcal{X} \times \mathcal{Y}$ be the product of μ and ν. Let $Z = X \times Y$. If the function $\psi : Z \to \mathbb{R}$ is integrable with respect to π, then the extended real-valued functions defined almost everywhere by

$$f(x) = \int_Y \psi_x\,d\nu \quad \text{and} \quad g(y) = \int_X \psi^y\,d\mu$$

have finite integrals and

$$\int_X f \, d\mu = \int_Z \psi \, d\pi = \int_Y g \, d\nu \, .$$

In other words,

$$\int_X \left(\int_Y \psi \, d\nu \right) d\mu = \int_Z \psi \, d\pi = \int_Y \left(\int_X \psi \, d\mu \right) d\nu \, .$$

Hahn decomposition The sets P and N in the Hahn decomposition theorem.

Hahn decomposition theorem If λ is a signed measure on the σ-algebra $\mathcal{X}$ on the set X, then there exist sets P and N in $\mathcal{X}$ with $X = P \cup N$, $P \cap N = \emptyset$, and such that P is positive and N is negative with respect to λ.

Hahn Extension Theorem Assume that μ is a σ-finite measure on an algebra $\mathcal{A}$. Then there exists a unique extension $\widehat{\mu}$ of μ to a measure on a σ-algebra $\widehat{\mathcal{A}}$.

Heine-Borel theorem A set in $\mathbb{R}$ is compact if and only if it is closed and bounded.

Hölder's inequality The inequality

$$\|f \cdot g\|_{L^1} \leq \|f\|_{L^p} \cdot \|g\|_{L^q} \, .$$

for $1/p + 1/q = 1$.

homeomorphism An equivalence of topological spaces.

infimum of a set The greatest lower bound of a set.

inner measure Suppose that $E \subseteq I_n$ for some n. Define the inner measure $m_*(E)$ of E to be
$$m_*(E) = m(I_n) - m^*(I_n \setminus E) \, .$$

inner regularity Approximation of a Lebesgue measurable set from the inside by a compact set.

integrable function A function f is integrable if both its positive and negative parts are integrable.

integral of a simple function   If $s(x) = \sum_{j=1}^{k} a_j \chi_{E_j}$, then the integral of s is defined to be $\sum_{j=1}^{k} a_j \mu(E_j)$.

Lebesgue Decomposition Theorem   Let λ and μ be σ-finite measures defined on a σ-algebra $\mathcal{X}$. There exists a measure λ_1 which is singular with respect to μ and another measure λ_2 which is absolutely continuous with respect to μ so that $\lambda = \lambda_1 + \lambda_2$. The measures λ_1 and λ_2 are unique.

Lebesgue dominated convergence theorem   If $f_j \to f$ pointwise and if $|f_j| \le g$ for some integrable function g and for all j, then

$$\lim_{j \to \infty} \int f_j \, d\mu = \int \lim_{j \to \infty} f_j \, d\mu \,.$$

Lebesgue integral   A theory of the integral in which the integral of a non-negative function f is the supremum of integrals of simple functions which lie below f.

Lebesgue measure   A measure which extends the length measure on intervals to the entire σ-algebra of Borel sets.

Lebesgue monotone convergence theorem   If $0 \le f_1 \le f_2 \le \ldots$ then

$$\lim_{j \to \infty} \int f_j \, d\mu = \int \lim_{j \to \infty} f_j \, d\mu \,.$$

Lebesgue null set   A set $E \subseteq \mathbb{R}$ is called a *Lebesgue null set* if $m^*(E) = 0$.

Lebesgue outer measure   Let $E \subseteq \mathbb{R}$. We define the Lebesgue outer measure $m^*(E)$ of E to be

$$m^*(E) = \inf \left\{ \sum_{j=1}^{\infty} \ell(I_j) \right\} \,,$$

where the infimum is taken over all sequences $\{I_j\}$ of open intervals in $\mathbb{R}$ that cover E in the sense that

$$E \subseteq \bigcup_{j=1}^{\infty} I_j \,.$$

Here of course $\ell(I_j)$ denotes the ordinary length of the interval I_j.

Lebesgue σ-algebra If m^* is Lebesgue outer measure, then the σ-algebra $\mathcal{L}$ of subsets of $\mathbb{R}$ that satisfies the Carathéodory condition is called the Lebesgue σ-algebra of $\mathbb{R}$. A set $E \in \mathcal{L}$ is called a Lebesgue measurable subset of $\mathbb{R}$ or, briefly, a measurable subset of $\mathbb{R}$. The restriction of m^* to $\mathcal{L}$, which we now call m, is called the Lebesgue measure on $\mathbb{R}$.

Lebesgue spaces   Spaces of functions which are pth power integrable, $1 \leq p < \infty$. Or the space of functions which are essentially bounded. These are denoted L^p and L^∞ respectively.

Lebesgue-Stieltjes measure   The extension of Borel-Stieltjes measure to a complete σ-algebra which contains the Borel sets.

length of an interval   The length of the interval $[a, b]$ is $b - a$.

limit infimum of a sequence   The least limit of any subsequence of the sequence.

limit supremum of a sequence   The greatest limit of any subsequence of the sequence.

linear functional   A linear functional on a linear space V is a mapping $\varphi : V \to \mathbb{R}$ which is linear.

linear space   A vector space.
measurable function   Let $\mathcal{X}$ be a σ-algebra on $\mathbb{R}$. A function $f : \mathbb{R} \to \mathbb{R}$ is said to be $\mathcal{X}$-measurable if, for each real number α, the set

$$\{x \in \mathbb{R} : f(x) > \alpha\}$$

belongs to $\mathcal{X}$.

measurable set   A set which is an element of a certain σ-algebra. Alternatively, a set which satisfies Carathéodory's condition.

measure on an algebra   Let $\mathcal{A}$ be an algebra of subsets of a set X. A measure on $\mathcal{A}$ is a function $\mu : \mathcal{A} \to \mathbb{R}^+$ satisfying:

(a) $\mu(\emptyset) = 0$;
(b) $\mu(E) \geq 0$ for all $E \in \mathcal{A}$;

(c) If $\{E_j\}$ is a sequence of pairwise disjoint sets in $\mathcal{A}$ such that $\cup_{j=1}^{\infty} E_j$ also belongs to $\mathcal{A}$, then

$$\mu\left(\bigcup_{j=1}^{\infty} E_j\right) = \sum_{j=1}^{\infty} \mu(E_j).$$

measure space   An ordered pair $(X, \mathcal{X})$, where $\mathbb{R}$ is the real numbers and $\mathcal{X}$ is a σ-algebra. Alternatively, an ordered triple $(X, \mathcal{X}, \mu)$, where X is a set, $\mathcal{X}$ is a σ-algebra on X, and μ is a measure on $\mathcal{X}$.

measure zero   A set has measure zero if it can be covered by open intervals the sum of whose lengths is less than ϵ for any $\epsilon > 0$.

mesh of a partition   The maximal length of an interval in the partition:

$$m(\mathcal{P}) = \min_{j=1,\dots,k} \triangle_j.$$

Minkowski's inequality   The inequality

$$\|f + g\|_{L^p} \le \|f\|_{L^p} + \|g\|_{L^p},$$

for $1 \le p \le \infty$.

monotone class   A monotone class is a nonempty collection $\mathcal{M}$ of sets which contains the union of each increasing sequence in $\mathcal{M}$ and also the intersection of each decreasing sequence in $\mathcal{M}$.

Monotone Class Lemma   If $\mathcal{A}$ is an algebra of sets, then the σ-algebra $\mathcal{S}$ generated by $\mathcal{A}$ coincides with the monotone class $\mathcal{M}$ generated by $\mathcal{A}$.

monotonicity for integrals   If $f \le g$, then

$$\int f \, d\mu \le \int g \, d\mu.$$

Alternatively, if $E \subseteq F$ and $f \ge 0$, then

$$\int_E f \, d\mu \le \int_F f \, d\mu.$$

$\widehat{\mu}$ measurability   Let X be a set and $\mathcal{A}$ an algebra on X. Let $\widehat{\mu}$ be an outer measure. A subset E of X is said to be $\widehat{\mu}$-measurable if

$$\widehat{\mu}(A) = \widehat{\mu}(A \cap E) + \widehat{\mu}(A \setminus E)$$

for all subsets $A \subseteq X$. The collection of all $\widehat{\mu}$-measurable sets is denoted by $\widehat{\mathcal{A}}$.

negative variation for λ The negative variation for λ is the finite measure

$$\lambda^-(E) = -\lambda(E \cap N).$$

negative with respect to λ A set $N \in \mathcal{X}$ is said to be negative with respect to the measure λ if $\lambda(E \cap N) \leq 0$ for any $E \in \mathcal{X}$.

norm A real-valued function N on a vector space V is said to be a norm if

(a) $N(\mathbf{v}) \geq 0$ for all $\mathbf{v} \in V$.

(b) $N(\mathbf{v}) = 0$ if and only if $\mathbf{v} = 0$.

(c) $N(\alpha \mathbf{v}) = |\alpha| N(\mathbf{v})$ for all $\mathbf{v} \in V$ and all real α.

(d) $N(\mathbf{u} + \mathbf{v}) \leq N(\mathbf{u}) + N(\mathbf{v})$ for all $\mathbf{u}, \mathbf{v} \in V$.

normed linear space A vector space equipped with a norm.

null set A set of measure zero.

null set with respect to λ A set $M \in \mathcal{X}$ is said to be a null set for λ if $\lambda(E \cap M) = 0$ for any set $E \in \mathcal{X}$.

outer measure generated by μ Let X be a set and $\mathcal{A}$ an algebra on X. Let μ be a measure on an algebra $\mathcal{A}$. If F is an arbitrary subset of X, then we define

$$\widehat{\mu}(F) = \inf \sum_{j=1}^{\infty} \mu(E_j),$$

where the infimum is extended over all sequences $\{E_j\}$ of sets in $\mathcal{A}$ such that

$$F \subseteq \bigcup_{j=1}^{\infty} E_j.$$

outer regularity Approximation of a Lebesgue measurable set from the outside by an open set.

partition A sequence of points $\mathcal{P} = \{x_0, x_1, \ldots, x_k\}$ in the interval $[a, b]$ with

$$a = x_0 \leq x_1 \leq x_2 \leq \cdots \leq x_k = b.$$

pointwise convergence The sequence $\{f_j\}$ of functions converges pointwise to f if $\lim_{j\to\infty} f_j(x)$ exists for each x.

positive variation for λ Let λ be a signed measure on $\mathcal{X}$ and let P, N be a Hahn decomposition for λ. The positive variation for λ is the finite measure

$$\lambda^+(E) = \lambda(E \cap P).$$

positive with respect to λ A set $P \in \mathcal{X}$ is said to be positive with respect to λ if $\lambda(E \cap P) \geq 0$ for any $E \in \mathcal{X}$.

product measure Given measure spaces $(X, \mathcal{X}, \mu)$ and $(Y, \mathcal{Y}, \nu)$, one constructs measurable rectangles and uses them to generate a σ-algebra. The product measure π has the property that

$$\pi(A \times B) = \mu(A) \cdot \nu(B).$$

Radon-Nikodým derivative The function f whose existence is establishd in the Radon-Nikodým theorem.

Radon-Nikodým theorem Let λ and μ be σ-finite measures defined on a σ-algebra $\mathcal{X}$. Suppose that λ is absolutely continuous with respect to μ. Then there is a measurable function f on $\mathcal{X}$ such that

$$\lambda(E) = \int_E f \, d\mu \quad \text{for all } E \in \mathcal{X}.$$

The function f is uniquely determined almost everywhere.

rectangle Let $(X, \mathcal{X})$ and $(Y, \mathcal{Y})$ be measure spaces. Then a set of the form $A \times B$, with $A \in \mathcal{X}$ and $Y \in \mathcal{Y}$ is called a measurable rectangle, or sometimes simply a rectangle in $Z \equiv X \times Y$.

Riemann integral The limit of the Riemann sums as the mesh of the partition goes to 0.

Riemann sum The quantity

$$\mathcal{R}_{\mathcal{P}} = \sum_{j=1}^{k} f(\xi_j) \cdot \triangle_j,$$

where ξ_j is an element of the interval I_j of the partition.

Riesz Representation Theorem, first version   If $(X, \mathcal{X}, \mu)$ is a σ-finite measure space and φ is a bounded linear functional on L^1, then there exists a function $g \in L^\infty$ such that equation

$$\varphi(f) = \int fg \, d\mu$$

holds for all $f \in L^1$. Furthermore, $\|\varphi\| = \|g\|_{L^\infty}$. Also $g \geq 0$ if φ is a positive linear functional.

Riesz Representation Theorem, second version   Let $(X, \mathcal{X}, \mu)$ be an arbitrary measure space. Let φ be a bounded linear functional on L^p, $1 < p < \infty$. Then there exists a $g \in L^q$, $q = p/(p-1)$, so that

$$\varphi(f) = \int fg \, d\mu$$

holds for all $f \in L^p$. Moreover, $\|\varphi\| = \|g\|_{L^q}$.

Riesz Representation Theorem, third version   Let $I = [0, 1]$. Let φ be a positive, bounded linear functional on $C([0, 1])$. Then there exists a measure γ defined on the Borel subsets of I such that

$$\varphi(f) = \int_I f \, d\gamma$$

for all $f \in C(I)$. Furthermore, the norm $\|\varphi\|$ of φ equals $\gamma(I)$.

seminorm   Like a norm, but failing property **(b)** of the definition of norm.

σ-algebra   A collection of sets which satisfies certain set-theoretic closure properties.

σ-finite measure   A measure which allows the space X to be written as the countable union of subsets X_j and each X_j has finite measure.

signed measure   A measure which takes both positive and negative values. It does not take the values $\pm\infty$.

simple function   A linear combination of characteristic functions.

singularity of measures   Two measures λ and μ on a σ-algebra $\mathcal{X}$ are said to be mutually singular if there are disjoint sets $A, B \in \mathcal{X}$ so that $x = A \cup B$

and $\lambda(A) = \mu(B) = 0$. In these circumstances we write $\lambda \perp \mu$.

supremum of a set   The least upper bound of the set.

Tonelli's Theorem   Let $(X, \mathcal{X}, \mu)$ and $(Y, \mathcal{Y}, \nu)$ be σ-finite measure spaces. Let $Z = X \times Y$ and let $\psi : Z \to \widehat{\mathbb{R}}$ be measurable and nonnegative. Then the functions can be defined on X and Y by

$$f(x) = \int_Y \psi_x \, d\nu \quad \text{and} \quad g(y) = \int_X \psi^y \, d\mu$$

are measurable and

$$\int_X f \, d\mu = \int_Z \psi d\pi = \int_Y g \, d\nu \, .$$

total variation for λ   The measure $|\lambda|$ which is defined for $E \in \mathcal{X}$ by

$$|\lambda|(E) = \lambda^+(E) + \lambda^-(E) \, .$$

translate of a set   If E is a set and $a \in \mathbb{R}$, then the translate of E by a is $E_a = \{e + a : e \in E\}$.

triangle inequality   The inequality

$$N(\mathbf{u} + \mathbf{v}) \le N(\mathbf{u}) + N(\mathbf{v})$$

for a norm N.

uncountable set   A set with cardinality greater than that of the natural numbers.

uniform convergence   A sequence of functions $\{f_j\}$ converges uniformly to f if, given $\epsilon > 0$, there is a $J > 0$ such that, for $j > J$, $|f_j(x) - f(x)| < \epsilon$.

vector space   A set V equipped with operations of addition and scalar multiplication and satisfying certain axioms.

x-section of E   Given a set $E \subseteq X \times Y$ and $x \in X$, this is the set

$$E_x = \{y \in Y : (x, y) \in E\} \, .$$

x-section of f   Let $Z = X \times Y$. If $f : Z \to \widehat{\mathbb{R}}$ and if $x \in X$, then the x-section of f is the function f_x defined on Y by

$$f_x(y) = f(x, y) \, , \quad y \in Y \, .$$

y-section of E   Given a set $E \subseteq X \times Y$ and $y \in Y$, this is the set

$$E^y = \{x \in X : (x, y) \in E\}.$$

y-section of f   Let $Z = X \times Y$. If $f : Z \to \widehat{\mathbb{R}}$ and if $y \in Y$, then the y-section of f is the function f^y defined on X by

$$f^y(x) = f(x, y), \quad x \in X.$$

$\mathcal{Z}$-measurable set   If $(X, \mathcal{X})$ and $(Y, \mathcal{Y})$ are measure spaces, then $\mathcal{Z} = \mathcal{X} \times \mathcal{Y}$ denotes the σ-algebra of subsets of $Z = X \times Y$ generated by rectangles $A \times B$ with $A \in \mathcal{X}$ and $B \in \mathcal{Y}$. A set in $\mathcal{Z}$ is called a $\mathcal{Z}$-measurable set, or sometimes just a measurable subset of Z.

Solutions to Selected Exercises

Chapter 1

1. Let $x \in [a, b]$ and $j > 0$. Then $a - 1/j < x < b + 1/j$. So $x \in (a - 1/j, b + 1/j)$. We conclude that $x \in \cap_{j=1}^{\infty}(a - 1/j, b + 1/j)$.

   Conversely, let $x \in \cap_{j=1}^{\infty}(a - 1/j, b + 1/j)$. Then $a - 1/j < x < b + 1/j$ for $j = 1, 2, \ldots$. If $x < a$ and $j > a - x$ then $a - 1/j \not< x$. If $x > b$ and $j > x - b$ then $x \not< b + 1/j$. We conclude that $a \leq x \leq b$. So $x \in [a, b]$.

   Putting the two results together gives

$$[a, b] = \bigcap_{j=1}^{\infty}(a - 1/j, b + 1/j).$$

   The second identity is proved similarly.

3. Suppose that x belongs to infinitely many of the sets B_n. Then $x \in B_{n_1}$, $x \in B_{n_2}$, ... with $n_1 < n_2 < \cdots$. But then $x \in \cup_{j=k}^{\infty} B_j$ for any k. In conclusion, $x \in \cap_{k=1}^{\infty} \cup j = k^{\infty} B_j$.

   Conversely, suppose that $x \in \cap_{k=1}^{\infty} \cup j = k^{\infty} B_j$. Then, for each k, $x \in \cup_{j=k}^{\infty} B_j$. Thus, for each k, there is a $j_k \geq k$ with $x \in B_{j_k}$. But this says that x lies in infinitely many of the B_j.

5. Let S be a set that is not Borel. The Lebesgue non-measurable set that we constructed is such a set. Define

$$f(x) = \begin{cases} 1 & \text{if} \quad x \in S, \\ -1 & \text{if} \quad x \notin S. \end{cases}$$

   Then f is not Borel measurable. But $f^2(x) \equiv 1$ is Borel measurable. And $|f|(x) \equiv 1$ is Borel measurable.

7. The function f differs from the function s_k by 2^{-k} on $S_{j,k}$.

9. If the inverse image of any Borel set lies in $\mathcal{X}$ then the image of any open set lies in $\mathcal{X}$. So f is measurable.

Conversely, if f is measurable, then the inverse image of any open set lies in $\mathcal{X}$. But this property is closed under countable intersections and unions, so the inverse image of any Borel set lies in $\mathcal{X}$.

11. (a) The empty set is denumerable. The complement of $\mathbb{R}$ is $\emptyset$, and that set is denumerable.

   (b) If A is denumerable then $\mathbb{R} \setminus A$ has denumerable complement. If ^cA is denumerable, then $\mathbb{R} \setminus {}^cA = A$ has denumerable complement.

   (c) If each a_j lies in $\mathcal{X}$ and if any of the a_j has denumerable complement, then $\cup_j a_j$ has an even smaller complement; so it, too, is denumerable. If none of the a_j has denumerable complement then each of the a_j is denumerable hence $\cup_j a_j$ is denumerable.

Chapter 2

1. (a) $\lambda(\emptyset) = \mu(K \cap \emptyset) = \mu(\emptyset) = 0$.

   (b) Let $E_1, E_2, \ldots$ be pairwise disjoint sets in $\mathcal{X}$. Then

$$
\lambda \left(\bigcup_{j=1}^{\infty} E_j \right) = \mu \left(K \cap \bigcup_{j=1}^{\infty} E_j \right)
$$
$$
= \mu \left(\bigcup_{j=1}^{\infty} (K \cap E_j) \right)
$$
$$
= \bigcup_{j=1}^{\infty} \mu(K \cap E_j)
$$
$$
= \bigcup_{j=1}^{\infty} \lambda(E_j).
$$

3. (a) The set $\emptyset$ is denumerable. So $\mu(\emptyset) = 0$.

   (b) Let $E_1, E_2, \ldots$ be pairwise disjoint sets in $\mathcal{X}$. If each of the D_j is denumerable then $\cup_j$ is denumerable and we have

$$
0 = \mu \left(\bigcup_{j=1}^{\infty} E_j \right) = \sum_{j=1}^{\infty} \mu(E_j).
$$

   If instead at least one of the E_j is uncountable, say E_{j_0}, then $\mu(E_{j_0}) = +\infty$ and $\mu(\cup_j E_j) = +\infty$. So the required identity is still true.

5. By Example 2.2, part **(d)**, the measure of the interval $(p - \epsilon, p + \epsilon)$ is 2ϵ for any $\epsilon > 0$. Thus $\mu(\{p\}) \leq 2\epsilon$ for every ϵ. We conclude that $\mu(\{p\}) = 0$.

Let $E = \{p_j\}$ be a countable set. Then

$$E \subseteq \bigcup_{j=1}^{\infty} \left(p_j - \epsilon/2^{j+1}, p_j + \epsilon/2^{j+1} \right) .$$

It follows that $\mu(E) \leq \sum_j \epsilon/2^j = \epsilon$ for any $\epsilon > 0$. Thus $\mu(E) = 0$.

For any small $\epsilon > 0$,

$$[a + \epsilon, b - \epsilon] \subseteq (a, b) \subseteq [a - \epsilon, b + \epsilon] .$$

Thus

$$(b - a) - 2\epsilon \leq \mu((a, b)) \leq (b - a) + 2\epsilon .$$

It follows that $\mu((a, b)) = b - a$.

The result for the other intervals is proved similarly.

7. From the unit interval we remove one interval of length $1/3$, two intervals of length $1/9$, four intervals of length $1/27$, and so forth. Thus the sum of the lengths of all the removed intervals is

$$\sum_{j=0}^{\infty} \frac{2^j}{3^{j+1}} = \frac{1}{3} \sum_{j=0}^{\infty} \left(\frac{2}{3} \right)^j = \frac{1}{3} \cdot \frac{1}{1 - 2/3} = 1 .$$

We conclude that the Lebesgue measure of the Cantor set is equal to the Lebesgue measure of the unit interval less the length of the union of the removed intervals. This is $1 - 1 = 0$.

9. Let us begin with the irrational numbers in the unit interval. The set of rational numbers in the unit interval, being a countable set, has measure 0. Thus the set of irrational numbers in the unit interval has measure 1. The same reasoning applies to the set of irrational numbers in any interval of the form $[j, j+1]$. Therefore the measure of the set of irrational numbers in the real line is $+\infty$.

11. We see that
$$\lambda(\emptyset) = \mu(f^{-1}(\emptyset)) = \mu(\emptyset) = 0 .$$

Also, if $E_1, E_2, \ldots$ are pairwise disjoint Borel sets then

$$\lambda \left(\bigcup_{j=1}^{\infty} E_j \right) = \mu \left(f^{-1} \left(\bigcup_{j=1}^{\infty} E_j \right) \right) = \sum_{j=1}^{\infty} \mu \left(f^{-1}(E_j) \right) = \sum_{j=1}^{\infty} \lambda(E_j) .$$

These two results show that λ is a Borel measure.

Chapter 3

1. Let $s_1(x) = \sum_{j=1}^{k} a_j \chi_{E_j}$ and $s_2(x) = \sum_{\ell=1}^{m} b_\ell \chi_{F_\ell}$ be simple functions. Then

$$s_1(x) + s_2(x) \quad = \quad \sum_{j=1}^{k}\sum_{\ell=1}^{m}(a_j + b_\ell)\chi_{a_j \cap b_\ell}(x)$$

$$+ \sum_{j=1}^{k} a_j\, \chi_{a_j \setminus \cup_{\ell=1}^{m} b_\ell}(x) + \sum_{\ell=1}^{m} b_\ell\, \chi_{b_\ell \setminus \cup_{j=1}^{k} a_j}(x).$$

This is clearly another simple function.

Likewise

$$s_1(x) \cdot s_2(x) = \sum_{j=1}^{k}\sum_{\ell=1}^{m} a_j \cdot b_\ell\, \chi_{a_j \cap b_\ell}(x).$$

This is clearly another simple function.

If c is a scalar then

$$c \cdot s_1(x) = \sum_{j=1}^{k} c a_j\, \chi_{E_j}(x).$$

This last is a simple function.

3. Define

$$s_m(x) = \sum_{\ell=1}^{m} a_\ell \chi_{[\ell, \ell+1)}.$$

Then clearly $s_1 \leq s_2 \leq \cdots$ and $s_j \to f$. Also

$$\int s_m(x)\, d\mu(x) = \sum_{\ell=1}^{m} a_\ell.$$

It follows therefore that

$$\int f(x)\, d\mu(x) = \sum_{\ell=1}^{\infty} a_\ell.$$

5. Certainly

$$f_{j+1}(x) = (1/(j+1))\chi_{[(j+1),+\infty)} \leq (1/j)\chi_{[j,+\infty)}.$$

So the sequence of functions is monotonically decreasing.

Let $\epsilon > 0$. Choose j so large that $j > 1/\epsilon$. Then

$$|f_j(x) - 0| \le 1/j < \epsilon$$

for every x. So the f_j converge uniformly to 0.

Finally, it is clear that $\int 0 \, d\mu(x) = 0$ and $\int f_j(x) \, d\mu(x) = +\infty$.

7. The sequence $\{g_j\}$ does not converge uniformly to 0 because $g_j(3/(2j)) = j$. Also the sequence is not monotonic.

But Fatou's lemma does apply. We see that

$$\int \liminf_{j \to \infty} g_j(x) \, d\mu(x) = \int 0 \, d\mu(x) = 0$$

while

$$\liminf_{j \to \infty} \int g_j(x) \, d\mu(x) = \liminf_{j \to \infty} \int 1 \, d\mu(x) = 1 \,.$$

9. Suppose that $f(x) = \sum_{j=1}^{k} a_j \chi_{I_j}(x)$, where the I_j are pairwise disjoint, closed intervals. Then of course

$$\int f(x) \, d\mu(x) = \sum_{j=1}^{k} a_j |I_j| \,,$$

where $|I_j|$ denotes the length of the interval I_j.

On the other hand, if a is the leftmost endpoint of any interval I_j and b is the rightmost endpoint of any interval I_j, then

$$
\begin{aligned}
\int_a^b f(x) \, dx &= \sum_{j=1}^{k} \int_{I_j} f(x) \, dx \\
&= \sum_{j=1}^{k} \int_{\alpha_j}^{\beta_j} f(x) \, dx \\
&= \sum_{j=1}^{k} \int_{\alpha_j}^{\beta_j} a_j \, dx \\
&= \sum_{j=1}^{k} a_j |I_j| \,,
\end{aligned}
$$

where $I_j = [\alpha_j, \beta_j]$.

11. Write

$$Z = \{x \in X : f(x) > 0\} = \bigcup_{j=1}^{\infty} \{x \in X : f(x) > 1/j\}.$$

But

$$\mu(\{x \in X : f(x) > 1/j\}) = \int_{\{x \in X : f(x) > 1/j\}} 1 \, d\mu(x)$$

$$\leq \int_{\{x \in X : f(x) > 1/j\}} \frac{f(x)}{1/j} \, d\mu(x)$$

$$< +\infty.$$

This shows that Z is σ-finite.

Chapter 4

1. Write

$$\mu(\{x \in X : |f(x)| > \alpha\}) = \int_{\{x \in X : |f(x)| > \alpha\}} 1 \, d\mu(x)$$

$$\leq \int_{\{x \in X : |f(x)| > \alpha\}} \frac{|f(x)|}{\alpha} \, d\mu(x)$$

$$< +\infty.$$

Furthermore,

$$\{x \in X : f(x) \neq 0\} = \bigcup_{j=1}^{\infty} \{x \in X : |f(x)| > 1/j\}.$$

So $\{x \in X : f(x) \neq 0\}$ is σ-finite.

3. Let h be integrable and let E be a measurable set. Assume without loss of generality that $h \geq 0$. Let s_j be simple functions that increase to h. Then $s_j \cdot \chi_E$ are simple functions that increase to $h\chi_E$. So $h\chi_E$ is integrable.

If g is bounded, measurable then g is integrable. Assume that $g \geq 0$. Let t_j be simple functions that increase to g. Then $t_j \cdot f$ is integrable. Passing to the limit, we see that $g \cdot f$ is integrable.

5. Equivalently, let us show that if g is integrable and $\int_E g \, d\mu(x) = 0$ for every set $E \in \mathcal{X}$, then $g \equiv 0$.

Suppose not. Then there is a set $F \in \mathcal{X}$ on which g is not zero. Passing to a subset, we may suppose that g is positive on F. But then $\int_F g \, d\mu(x) > 0$. That is a contradiction.

7. Let $f_j(x) \equiv 1/j$ on $\mathbb{R}$. Then $f_j \to 0 \equiv f$ uniformly but $\int f_j \, d\mu(x) \not\to \int f \, d\mu(x)$.

9. There is a subsequence $j_1, j_2, \ldots$ such that

$$\sum_{\ell=j_k}^{j_{k+1}} \|f_\ell\|_{L^1} < 2^{-k}.$$

Let

$$g_k = \sum_{\ell=j_1}^{j_k} |f_\ell| \qquad \text{and} \qquad g = \sum_{\ell=j_1}^{\infty} |f_\ell|.$$

Certainly $\|g_k\|_{L^1} < 1$ for each k. Fatou's lemma applied to $\{g_k\}$ then yields that $\|g\|_{L^1} \le 1$. It follows that $|g(x)| < +\infty$ for almost every x. As a result,

$$f_{j_1}(x) + \sum_{\ell=1}^{\infty}(f_{j_{\ell+1}} - f_{j_\ell})$$

converges absolutely for almost every x.

11. Either use Riemann sums or integrate by parts.

Chapter 5

1. Clearly $C([0,1])$ is closed under addition and scalar multiplication. If $\{f_j\} \subseteq C([0,1])$ is a Cauchy sequence, then $\{f_j\}$ converges uniformly. So there is a continuous limit function f. Hence $C([0,1])$ is complete. We conclude that $C([0,1])$ is a Banach space.

3. Define

$$f_j(x) = \begin{cases} 0 & \text{if} & x \le 0, \\ jx & \text{if} & 0 < x \le 1/j, \\ 1 & \text{if} & 1/j < x < \infty. \end{cases}$$

Then the f_j converge in the $\| \ \|_1$ norm to

$$f(x) = \begin{cases} 0 & \text{if} & x \le 0, \\ 1 & \text{if} & x > 0. \end{cases}$$

This f is not continuous. Therefore the space is not complete in this norm so it is not a Banach space.

7. Define

$$f(x) = \frac{1}{x^{1/p_0}}.$$

Then

$$\int_0^1 |f(x)|^p \, d\mu(x) = \int_0^1 \frac{1}{x^{p/p_0}} \, d\mu(x).$$

We see that the integral converges when $p/p_0 < 1$ and diverges when $p \geq p_0$.

9. We see that

$$\int |f(x)|^r \, d\mu(x) = \int |f(x)|^r \cdot 1 \, d\mu(x)$$

$$\leq \int |f(x)|^p \, d\mu(x)^{r/p} \cdot \int 1 \, d\mu(x)^{(p-r)/p}$$

as long as $a \leq r \leq p$ by Hölder's inequality. That proves the result.

11. Certainly $f \cdot \chi_{[-j,j]} \to f$ in L^p norm as $j \to \infty$. Let $\epsilon > 0$ and choose j so large that $\|f - f \cdot \chi_{[-j,j]}\|_{L^p} < \epsilon$. Let $D = [-j,j]$. Now if $F \subseteq \mathbb{R}$ and $F \cap E = \emptyset$, then $\|f \cdot \chi_F\|_{L^p} < \epsilon$.

13. The triangle inequality for this norm is just Minkowski's inequality for the counting measure.

Chapter 6

2. Construct a Cantor-like set as follows:

    - Begin with the unit interval $I_0 = [0,1]$.
    - Remove the open middle interval with length $1/5$. This creates I_1.
    - Remove from each of the remaining intervals the open middle interval with length $1/25$. This creates I_2.
    - Remove from each of the remaining intervals the open middle interval with length $1/125$. This creates I_3.
    - Etcetera.

    The resulting set, obtained by intersecting all the I_j, is a compact, nonempty set. We can calculate its length by calculating the sum of the lengths of all the intervals removed. That number is

$$\sum_{j=0}^{\infty} \frac{2^j}{5^{j+1}} = \frac{1}{5} \cdot \sum_{j=0}^{\infty} \left(\frac{2}{5}\right)^j = \frac{1}{5} \cdot \frac{1}{1 - 2/5} = \frac{1}{3}.$$

    So the sum of the lengths of all the removed intervals is $1/3$. Thus the length of this new Cantor set is $2/3$.

5. If $E \subseteq \mathbb{R}$ has outer measure zero then, for each $\epsilon > 0$, there is a collection of intervals I_j so that $E \subseteq \cup_j I_j$ and $\sum_j \ell(I_j) < \epsilon$. But then $\sum_j \mu(I_j) < \epsilon$. And that says that $\mu(E) < \epsilon$. Since this is true for every $\epsilon > 0$, we conclude that $\mu(E) = 0$.

6. Let $E_1, E_2, \ldots$ each have outer measure 0. Let $\epsilon > 0$. Then, for each j, there are intervals $\{I_\ell^j\}$ so that $\sum_\ell \ell(I_\ell^j) < \epsilon/2^j$. But then

$$\{I_\ell^j\}_{j,\ell=1}^\infty$$

is a collection of intervals covering $\cup_j E_j$ and the sum of whose lengths does not exceed ϵ. Hence $\cup_j E_j$ has outer measure 0.

7. The Cantor ternary set is totally disconnected and has outer measure 0.

9. Let $E \subseteq \mathbb{R}$ be a set and let $\{I_j\}$ be a collection of intervals that covers E. For $a \in \mathbb{R}$, define $E_a = \{e+a : e \in E\}$. Then the sets $(I_j)_a$ cover E_a and the sum of the lengths of the $(I_j)_a$ is the same as the sum of the lengths of the I_j. That makes outer measure translation invariant.

Chapter 7

1. We saw in Lemma 7.6 that it is enough to show that if $A \subseteq \mathbb{R}$ is such that $m^*(A) < +\infty$, then

$$m^*(A) \geq m^*(A \cap I) + m^*(A \setminus I).$$

Let $n \in \mathbb{N}$ and let $I_n = \{x \in I : \mathrm{dist}(x, {}^c I) > 1/n\}$. Hence $I_n \subseteq I$. Also, since $I \setminus I_n$ lies in the union of 2 cells each of which has side length $1/n$, then $m^*(I \setminus I_n) \to 0$ as $n \to \infty$.

Notice that $A \supseteq (A \cap I_n) \cup (A \setminus I)$ and that $\mathrm{dist}(A \cap I_n, A \setminus I) \geq 1/n$. Thus we have from Proposition 6.6 that

$$\begin{aligned} m^*(A) &\geq m^*((A \cap I_n) \cup (A \setminus I)) \\ &= m^*(A \cap I_n) + m^*(A \setminus I). \end{aligned} \quad (7.10.1)$$

But we also know that

$$A \cap I = (A \cap I_n) \cup (A \cap (I \setminus I_n)).$$

Thus it follows from the subadditivity and monotone character of m^* that

$$m^*(A \cap I_n) \leq m^*(A \cap I) \leq m^*(A \cap I_n) + m^*(I \setminus I_n).$$

Thus we have

$$m^*(A \cap I) = \lim_{n \to \infty} m^*(A \cap I_n).$$

So, taking the limit in (7.10.1), we have

$$m^*(A) \geq m^*(A \cap I) + m^*(A \setminus I).$$

This shows, by Lemma 7.6, that I is a measurable set.

2. The complement of the Cantor set in the unit interval is an open set, which is measurable. Therefore the Cantor set is Lebesgue measurable.

3. The Cantor set has measure 0. Each C_j is a subset of the Cantor set hence also has measure 0. Thus the assertion is obvious.

4. The Cantor set has Lebesgue measure zero. And any translate of the Cantor set has Lebesgue measure zero.

5. Any interval of the form $[a, a + 1]$ has Lebesgue measure 1. And there are clearly uncountably many different possible values for a and hence uncountably many different such sets.

    For a slightly more interesting answer, let $p_1, p_2, \ldots, p_k$ be positive numbers that sum to 1. Let $E_{p_1, p_2, \ldots, p_k}$ be the disjoint union of k intervals $I_1, I_2, \ldots, I_k$ so that $\ell(I_j) = p_j$. Then clearly $\mu(E_{p_1, p_2, \ldots, p_k}) = 1$. There are clearly uncountably many sets of the form $E_{p_1, p_2, \ldots, p_k}$. So that does the trick.

8. Let $\mathcal{A}$ and $\mathcal{B}$ each be σ-algebras. Set $\mathcal{X} = \mathcal{A} \cap \mathcal{B}$. We claim that $\mathcal{X}$ is a σ-algebra.

   - Since $\emptyset \in \mathcal{A}$ and $\emptyset \in \mathcal{B}$, we see that $\emptyset \in \mathcal{X}$. Likewise, since $\mathbb{R} \in \mathcal{A}$ and $\mathbb{R} \in \mathcal{B}$, it follows that $\mathbb{R} \in \mathcal{X}$.
   - Let $E \in \mathcal{X}$. Then $E \in \mathcal{A}$. So $\mathbb{R} \setminus E \in \mathcal{A}$. Likewise, $E \in \mathcal{B}$. Hence $\mathbb{R} \setminus E \in \mathcal{B}$. Putting these two results together gives $\mathbb{R} \setminus E \in \mathcal{X}$.
   - If E_j is a sequence of sets in $\mathcal{X}$, then E_j is also a sequence of sets in $\mathcal{A}$ and in $\mathcal{B}$. So $\cup_j E_j \in \mathcal{A}$ and $\cup_j E_j \in \mathcal{B}$. It follows that $\cup_j E_j \in \mathcal{X}$.

   This establishes that $\mathcal{X}$ is a σ-algebra.

Chapter 8

1. Let P be a positive set. Let $E \in \mathcal{X}$ be any subset of P. Now let F be any element of $\mathcal{X}$. Then, by the definition of positivity, $P \cap (E \cap F)$ has nonnegative λ measure. But this says that $E \cap F$ has nonnegative λ measure. So E is positive.

3. By hypothesis, if $\mu_3(E) = 0$, then $\mu_2(E) = 0$. Also, if $\mu_2(E) = 0$, then $\mu_1(E) = 0$. As a result, if $\mu_3(E) = 0$, then $\mu_1(E) = 0$. Hence μ_1 is absolutely continuous with respect to μ_3.

4. Now
$$\lambda(\emptyset) = \sum_{j=1}^{\infty} 2^{-j} \mu_j(\emptyset) = \sum_{j=1}^{i} nfty0 = 0\,.$$

Also, if $E_1, E_2, \ldots$ are pairwise disjoint sets in $\mathcal{X}$, then

$$\lambda\left(\bigcup_{\ell} E_\ell\right) = \sum_{j=1}^{\infty} 2^{-j} \mu_j\left(\bigcup_{\ell} E_\ell\right) = \sum_{j=1}^{\infty}\sum_{\ell=1}^{\infty} \mu_j(E_\ell)$$

$$= \sum_{\ell=1}^{\infty}\sum_{j=1}^{\infty} \mu_j(E_\ell) = \sum_{\ell=1}^{\infty} \lambda(E_\ell)\,.$$

So λ is a measure.

If $\lambda(E) = 0$ then it must be that $\mu_j(E) = 0$ for each j. So μ_j is absolutely continuous with respect to λ for each j.

7. Since $\lambda \perp \mu$, there are disjoint sets A and B, with $A \cup B = X$, so that $\lambda(A) = 0$ and $\mu(B) = 0$. But if $\lambda \ll \mu$ then it follows that $\lambda(B) = 0$. Hence $\lambda(X) = 0$. So λ is the 0 measure.

8. Assume that $|\lambda| \perp \mu$. So there are disjoint sets A, B with $A \cup B = X$ so that $|lambda|(A) = 0$ and $\mu(B) = 0$. Now let P, N be a Hahn decomposition for λ. Let $\lambda^+(E) = \lambda(E \cap P)$ and $\lambda^-(E) = \lambda(E \cap N)$. Then
$$\lambda^+(A) = \lambda(A \cap P) = 0\,.$$

Likewise,
$$\lambda^-(A) = \lambda(A \cap N) = 0\,.$$

Hence λ^+ and λ^- are both singular with respect to μ.

10. Assume without loss of generality that $g \geq 0$. Let $q = p/(p-1)$. Set $f = g^{q-1}$. Then $f^p = g^q$. As a result,

$$\left|\int g^q \, d\mu\right| = \left|\int f \cdot g \, d\mu\right| = |\varphi(f)| \leq C \cdot \int |f|^p d\mu^{1/p} = C \cdot \int |g|^q \, d\mu^{1/p}\,.$$

This simplifies to

$$\left|\int g^q \, d\mu\right|^{1/q} \leq C\,.$$

So $g \in L^q$. We have shown that $\|g\|_{L^q} \leq \|\varphi\|$ and the opposite inequality is clear by inspection of the proof.

11. Define a measure λ by
$$\lambda = \delta_0 - \delta_1\,,$$

where δ represents the Diract delta mass at the indicated point. Then $P = (-\infty, 1/2]$, $N = (1/2, +\infty)$ is one Hahn decomposition for λ. And $P = (-\infty, 2/3]$, $N = (2/3, +\infty)$ is another Hahn decomposition for λ.

Chapter 9

1. This collection of sets is not an algebra because it is not closed under complementation. However, the σ-algebra generated by these four types of sets contains all open sets, all closed sets, and therefore all Borel sets.

2. Of course the set $(a, +\infty)$ has infinite measure. Therefore the measure or length of the set

$$\bigcup_{j=1}^{N} \ell((a_j, b_j]$$

increases without bound as $N \to +\infty$. This implies that

$$\sum_{j=1}^{\infty} \ell((a_j, b_j]) = +\infty\,.$$

3. It is clear that $\mathcal{A}$ is closed under complementation and finite union. The empty set belongs to $\mathcal{A}$ and the set of all rational numbers belongs to $\mathcal{A}$. So $\mathcal{A}$ is an algebra. Each nonempty set in $\mathcal{A}$ consists of all the rationals between a and b, where a is less than b. So such a set must be infinite.

5. The Cantor ternary set is such a set.

7. By the definition of outer measure, such a set U exists with ℓ^* replaced by ℓ. But the same set U will work for ℓ^*.

9. Again invoke the notion of outer measure to obtain the set U.

    For the second part, approximate E by an open set U. Then approximate the characteristic function of U by a piecewise linear function.

11. • Of course U contains a nontrivial interval, and the measure of that interval is positive.
     • Every compact set K is contained in a bounded interval, and the measure of that bounded interval is finite.
     • We have seen that outer measure is translation invariant. It follows that Lebesgue measure is translation invariant.

13. In fact all the cases are handled in just the same way.

Chapter 10

1. If $s \leq f$ is a simple function that approximates f, then $s(x + a)$ approximates $f(x + a)$.

2. Fubini's theorem shows that the operator makes good sense.

    Define the translation operator

    $$\tau_a f(x) = f(x - a)$$

    for $a \in \mathbb{R}$. Then

    $$\tau_a T f(x) \;=\; T f(x - a) = \int f(x - a - t)\varphi(t)\,dt$$

    $$=\; \int \tau_a f(x - t)\varphi(t)\,dt = T(\tau_a f)(x)\,.$$

3. This result follows immediately from Fubini's theorem.

4. Use Jensen's inequality.

5. This can be proved with a counting argument. On the other hand, if E is a Lebesgue measurable set that is not Borel, then χ_E is a Lebesgue measurable function that is not Borel measurable.

7. The nonmeasurable set cannot be Borel, because Borel sets are measurable.

Chapter 11

1. In fact one can simply replace the open intervals by their closures.

2. First, one can replace I_2 by $I_2 \setminus I_1$ and I_3 by $I_3 \setminus (I_1 \cup I_2)$ and so forth. In this way one gets pairwise disjoint intervals. Then one can replace each interval by a slightly smaller, open subinterval.

5. Fix the open set $\mathcal{O} \subseteq \mathbb{R}$. Define

    $$K_j = x \in \mathcal{O} : \mathrm{dist}(x, {}^c\mathcal{O}) \geq 1/j \cap \{x \in \mathbb{R} : |x| \leq j\}\,.$$

    Then K_j is clearly closed and bounded and $\cup_j K_j = \mathcal{O}$.

6. Let C be the Cantor ternary set. For $\epsilon > 0$, let

    $$U_\epsilon = \{x \in \mathbb{R} : \mathrm{dist}(x, C) < \epsilon\}\,.$$

    Plainly U_ϵ is open. And $\cap_{\epsilon > 0} U_\epsilon = C$.

7. Approximate ^cS from the outside by open sets. Then take the complement.

Chapter 12

1. If $x \in \mathbb{R}$, then
$$|f_j(x) - 0| \leq j^{-1/p} .$$

So the sequence of functions converges uniformly to 0.

However, if $k > j$, then

$$\int |f_k(x) - f_j(x)|^p \, \delta\mu(x) \geq \sum_{\ell=j}^{k} (\ell + 1)^{-1} .$$

The series on the right diverges, so the sequence $\{f_j\}$ is not Cauchy in L^p.

2. If $x > 0$, then there is a J so large that $2/J < x$. Thus $x \notin [1/j, 2/j]$ So, for $j > J$, $g_j(x) = 0$. All the g_j are equal to 0 at the origin.

Notice that

$$\int |g_j(x) - 0|^p \, d\mu(x) 1/p = \int_{1/j}^{2/j} j^p \, d\mu(x)^{1/p} = j^{(p-1)/p} .$$

This quantity tends to $+\infty$ with j.

3. First, for $\alpha > 0$,

$$\mu\{x : |f_j(x)| \geq \alpha\} = \mu\{x : j^{-1/p} \cdot \chi_{[0,j]}(x) \geq \alpha\} = 0$$

when j is large enough. So the f_j converge to 0 in measure.

Next, for $\alpha > 0$,

$$\mu\{x : |f_j(x)| \geq \alpha\} = \mu\{x : j \cdot \chi_{[1/j,2/j]}(x) \geq \alpha\} = 1/j$$

if $j > \alpha$. This expression tends to 0 as $j \to +\infty$.

4. Fix an $x \in \mathbb{R}$. If j is large enough then $h_j(x) = 0$. So the sequence $\{h_j\}$ converges to 0 at x. It does not converge in measure to 0 because, for $\alpha > 0$,

$$\mu\{x : |h_j(x)| \geq \alpha\} = 1$$

when $\alpha \leq 1$.

5. We have already shown that the sequence in Exercise 2 converges to 0 in measure. It does not converge in L^p because

$$\int |g_j(x) - 0|^p \, d\mu(x)^{1/p} = j^{(p-1)/p},$$

and this expression does not tend to 0 as $j \to +\infty$.

6. The subsequence consisting of

$$\chi_{[0,1/2]} \, , \, \chi_{[0,1/2^2]} \, , \, \chi_{[0,1/2^3]} \, \cdots$$

converges almost everywhere to 0. In fact it converges to zero at every point except the origin. Note that this subsequence converges to 1 at the origin. So this subsequence converges at every point.

7. A subsequence must converge pointwise. So it must converge to a characteristic function.

10. Let x be a point at which f_j converges. Let $\epsilon > 0$. Choose $\delta > 0$ so that $|\varphi(f(x)) - t| < \epsilon$ when $|f(x) - t| < \delta$. Now choose J so large that $j > J$ implies that $|f_j(x) - f(x)| < \delta$. Then, for $j > J$, we have

$$|\varphi(f_j(x)) - \varphi(f(x))| < \epsilon.$$

So $\varphi \circ f_j$ converges at x.

11. We calculate that

$$\int |\varphi \circ f(x)|^p \, d\mu(x)^{1/p} \; \le \; \int_{\{x : |f(x)| \le K\}} |\varphi \circ f(x)|^p \, d\mu(x)^{1/p}$$

$$+ \int_{\{x : |f(x)| \ge K\}} |\varphi \circ f(x)|^p \, d\mu(x)^{1/p}$$

$$\equiv \;\; I + II.$$

Since φ is continuous, it is bounded on $[-K, K]$ by some number M. Thus

$$I \le \int_{\{x : |f(x)| \le K\}} M^p \, d\mu(x)^{1/p} \le [\mu(X) M^p]^{1/p}.$$

We estimate the second integral by

$$II \le \int_{\{x : |f(x)| \ge K\}} |K f(t)|^p \, d\mu(x)^{1/p} \le K \cdot \int_X |f|^p \, d\mu^{1/p} = K \|f\|_{L^p}.$$

Chapter 13

1. If either A or B is empty then clearly $A \times B$ is empty. Conversely, if A and B are both nonempty then $A \times B$ is nonempty. So $A \times B$ empty implies that either A or B is empty.

2. Assume that $A_1 \times B_1 = A_2 \times B_2$. Let $a \in A_1$. If b is an element of B_1 then $(a, b) \in A_2 \times B_1$. So $(a, b) \in A_2 \times B_2$. Hence $a \in A_2$. Hence $A_1 \subseteq A_2$. A similar argument shows that $A_2 \subseteq A_1$. We conclude that $A_1 = A_2$.

   The same reasoning shows that $B_1 = B_2$.

3. Let $A = [a_1, a_2]$ and $B = [b_1, b_2]$. Then

$$\mathbb{R} \setminus A = (-\infty, a_1) \cup (a_2, +\infty) \equiv A_1 \cup A_2$$

   and

$$\mathbb{R} \setminus B = (-\infty, b_1) \cup (b_2, +\infty) \equiv B_1 \cup B_2.$$

   It follows that

$$(\mathbb{R} \times \mathbb{R}) \setminus (A \times B) = (\mathbb{R} \times B_2) \cup (A_1 \times B \cup A_2 \times B) \cup (\mathbb{R} \times B_1).$$

4. Let us do the problem for $k = 2$. Then

$$
\begin{aligned}
(A_1 \times B_1) \cup (A_2 \times B_2) \;=\; & (A_1 \setminus A_2) \times B_1 \cup (A_2 \setminus A_1) \times (B_1 \setminus B_2) \\
& \cup (A_1 \cap A_2) \times (B_1 \cap B_2) \\
& \cup (A_1 \cap A_2) \times (B_2 \setminus B_1) \\
& \cup (A_2 \setminus A_1) \times B_2.
\end{aligned}
$$

6. Let U be an open set in $\mathbb{R}$. We calculate that

$$
\begin{aligned}
h^{-1}(U) \;=\; & \{(x, y) \in \mathcal{X} \times \mathcal{Y} : h(x, y) \in U\} \\
=\; & \{(x, y) \in \mathcal{X} \times \mathcal{Y} : f(x) \cdot g(y) \in U\} \\
=\; & \bigcup_{x \in X} \{(x, y) : g(y) \in U/[f(x)]\}.
\end{aligned}
$$

   But, for fixed x, $\{y : g(y) \in U/[f(x)]\}$ is Borel. So $\{(x, y) : g(y) \in U/[f(x)]\}$ is Borel. And the union of Borel sets is Borel. Hence $h^{-1}(U)$ is Borel.

8. Let $x \in X$ and $y \in (E \setminus F)_x$. Then $(x, y) \in E \setminus F$. Thus $(x, y) \in E$ so $y \in E_x$. Also $(x, y) \notin F$. So $y \notin F_x$. It follows that $y \in E_x \setminus F_x$. For the converse direction, let $y \in E_x \setminus F_x$. Thus $y \in E_x$, hence

$(x, y) \in E$. Also $y \notin F_x$ so $(x, y) \notin F$. Thus $(x, y) \in E \setminus F$. So $y \in (E \setminus F)_x$.

Now let $y \in (\cup_j E_j)_x$. So $(x, y) \in \cup_j E_j$. Therefore $(x, y) \in E_j$ for some particular j. So $y \in (E_j)_x$. Hence $y \in \cup_j (E_j)_x$. So $(\cup_j E_j)_x \subseteq \cup_j (E_j)_x$. The converse direction is similar.

9. This is just Fubini's theorem for the counting measure.

10. For fixed j,

$$\sum_{k=1}^{\infty} a_{j,k} = a_{k,k} + a_{k,k_1} = 1 - 1 = 0.$$

It follows that

$$\sum_{j=1}^{\infty} \sum_{k=1}^{\infty} a_{j,k} = 0.$$

Next, for fixed k,

$$\sum_{j=1}^{\infty} a_{j,k} = a_{k,k} + a_{k-1,k} = 1 - 1 = 0$$

as long as $k \geq 2$. For $k = 1$,

$$\sum_{j=1}^{\infty} a_{j,k} = a_{k,k} = 1.$$

The reason that these sums in different orders give different answers is that the function

$$(j, k) \longmapsto a_{j,k}$$

is not integrable. In fact

$$\sum_{j,k} |a_{j,k}| = +\infty.$$

Chapter 14

1. Refer to the solution of Exercise 2 in Chapter 6.

3. Let $E \subseteq \mathbb{R}$ be a null set. Write

$$E = \bigcup_{j=-\infty}^{\infty} E \cap [j, j+1].$$

Then each $E \cap [j, j+1]$ is a null set and E is their union.

4. Write
$$[0,1] = \bigcup_{x \in [0,1]} \{x\} \,.$$

5. Let C be the Cantor ternary set. Let $I = [0,1]$ be the unit interval. Write
$$C = I \setminus (I \setminus C) \,.$$
We know from our earlier work that $m(I) = 1$ and $m(I \setminus C) = 1$. Thus
$$m^*(C) = 1 - 1 = 0 \,.$$

6. Let I be the unit interval and write
$$I = I \setminus (I \setminus I) = I \setminus \emptyset \,.$$
Of course $m(I) = 1$ and $m(\emptyset) = 0$. So
$$m^*(I) = 1 - 0 = 1 \,.$$

Chapter 15

1. Let $E \subseteq \mathbb{R}$ be Lebesgue measurable but not Borel. Then $E \times E$ is Lebesgue product measurable but not product Borel.

2. It is a classical result that $C + C = [0,2]$. We shall explain briefly why $C \ominus C = [-1,1]$. We examine the sets that are used, by way of intersection, to construct C. First note that
$$I_0 \ominus I_0 = [0,1] \ominus [0,1] = [-1,1] \,.$$
Next note that
$$I_1 \ominus I_1 = ([0,1/3] \cup [1/3,2/3]) \ominus ([0,1/3] \cup [1/3,2/3]) = [-1,1] \,.$$
Next note that
$$\begin{aligned} I_2 \ominus I_2 &= ([0,1/9] \cup [2/9,1/3] \cup [2/3,7,9] \cup [8/9,1]) \\ &\quad \ominus ([0,1/9] \cup [2/9,1/3] \cup [2/3,7,9] \cup [8/9,1]) \\ &= [-1,1] \,. \end{aligned}$$
The calculations proceed in the same way for all $I_j \ominus I_j$. Taking the intersection of the I_j, and passing to the limit in our calculations, we find that
$$C \ominus C = [-1,1] \,.$$

3. Of course the difference of two rational numbers is a rational number. And any rational number can be obtained as the difference of two other rational numbers. So $\mathbb{Q} \ominus \mathbb{Q} = \mathbb{Q}$.

   The difference of two irrational numbers can be either rational or irrational. Any real number can be obtained as the difference of two irrational numbers. So

$$(\mathbb{R} \setminus \mathbb{Q}) \ominus (\mathbb{R} \setminus \mathbb{Q}) = \mathbb{R}.$$

5. By the definition of the equivalence relation that was used to construct the nonmeasurable set, $S \ominus S$ will contain no rational numbers except 0. It is difficult to say anything more about $S \ominus S$.

7. Any subset of a null set is also a null set. So is still measurable.

8. We can apply the same argument to $S \cup E$ as we did to S to see that $S \cup E$ can have neither 0 measure nor positive measure. So $S \cup E$ is still nonmeasurable.

Chapter 16

2. If we replace $\varphi(x)$ with $\psi(x) = c \cdot \chi_{[-1,1]}(x) \cdot \varphi(x)$ then we can imitate the proof of Theorem 16.8 to get a result. But the difference between φ and ψ is a small tail which is easy to control.

3. This is because φ_ϵ converges to the Dirac delta measure in the topology of distributions. That is to say, if f is a C^∞ function with compact support, then

$$\lim_{\epsilon \to 0} \int f(x)\varphi_\epsilon(x)\,d\mu(x) = \int f(x)d\delta(x) = \delta(f) = f(0).$$

4. It suffices to prove a version of Theorem 16.7 for L^p. But for that it suffices to show that the Hardy-Littlewood maximal function is bounded on L^p. We know that M is weak-type $(1,1)$. And it is trivial that M is bounded on L^∞. Thus the Marcinkiewicz interpolation theorem (see [6]) yields L^p boundedness of M, $1 < p \leq \infty$.

5. The function
$$f(x) = x$$
   is locally integrable but not integrable.

6. Let f be locally integrable. Let $P \in \mathbb{R}$. Let φ be a continuous function with compact support centered at P so that $\varphi(x) = 1$ for $x \in (P-1, P+1)$. Then Theorems 16.7 and 16.8 both apply to the function $g \equiv \varphi \cdot f$. And the result that this yields for $x \in (P-1, P+1)$ is identical to the result for f on that same interval.

References

[1] R. G. Bartle. *The Elements of Integration and Lebesgue Measure*. Wiley, New York, 1966.

[2] V. G. Kanovei. A proof of a theorem of Lusin. *Mat. Zametki*, 23:61, 1978.

[3] S. G. Krantz. *The Elements of Advanced Mathematics*, 4th ed. CRC Press, Boca Raton, FL, 2017.

[4] S. G. Krantz. *Real Analysis and Foundations*, 4th ed. CRC Press, Boca Raton, FL, 2017.

[5] R. Solovay. A model of set theory in which every set is lebesgue measurable. *Annals of Mathematics*, 92:1–56, 1970.

[6] E. M. Stein and G. Weiss. *Introduction to Fourier Analysis on Euclidean Spaces*. Princeton University Press, Princeton, NJ, 1971.

[7] K. R. Stromberg. *An Introduction to Classical Real Analysis*. Wadsworth, Belmont, CA, 1981.

Index

Milton Keynes UK
Ingram Content Group UK Ltd.
UKHW030900141024
449569UK00025B/1309